Identification Discs of
Union Soldiers in the Civil War

Identification Discs of Union Soldiers in the Civil War

A Complete Classification Guide and Illustrated History

LARRY B. MAIER *and*
JOSEPH W. STAHL

Foreword by EDWIN C. BEARSS

McFarland & Company, Inc., Publishers
Jefferson, North Carolina, and London

The present work is a reprint of the illustrated case bound edition of Identification Discs of Union Soldiers in the Civil War: A Complete Classification Guide and Illustrated History, *first published in 2008 by McFarland.*

Frontispiece: An unknown soldier wearing an identification disc (courtesy U.S. Army Military History Institute)

LIBRARY OF CONGRESS CATALOGUING-IN-PUBLICATION DATA

Maier, Larry B., 1949–
Identification discs of Union soldiers in the Civil War : a complete classification guide and illustrated history / Larry B. Maier and Joseph W. Stahl ; foreword by Edwin C. Bearss.
p. cm.
Includes bibliographical references and index.

ISBN 978-0-7864-6106-6
soft cover : 50# alkaline paper ∞

1. United States—History—Civil War, 1861–1865—Equipment and supplies—Handbooks, manuals, etc. 2. Medals—United States—Handbooks, manuals, etc. 3. Badges—United States—Handbooks, manuals, etc. 4. Soldiers—United States—Identification—Handbooks, manuals, etc. 5. Dead—Identification—Handbooks, manuals, etc. 6. United States. Army—Equipment—Handbooks, manuals, etc. 7. Soldiers—United States—Identification—History—19th century—Miscellanea. 8. United States. Army—History—Civil War, 1861–1865—Miscellanea. 9. United States—History—Civil War, 1861–1865—Miscellanea. I. Stahl, Joseph W., 1946– II. Title.
E491.M13 2010   973.7'6—dc22   2008023126

British Library cataloguing data are available

©2008 Larry B. Maier and Joseph W. Stahl. All rights reserved

No part of this book may be reproduced or transmitted in any form or by any means, electronic or mechanical, including photocopying or recording, or by any information storage and retrieval system, without permission in writing from the publisher.

On the cover: Union soldiers' identification discs of various types and condition (from the authors' collection; photograph by Joseph W. Stahl)

Manufactured in the United States of America

*McFarland & Company, Inc., Publishers
Box 611, Jefferson, North Carolina 28640
www.mcfarlandpub.com*

To the soldiers whose identification discs remind us
of their lives, their service, and their sacrifice

Acknowledgments

Many people think about writing a book. Often it is the ones who are blessed with the support and assistance of others who actually manage to get it accomplished. For me, this was especially true since I lacked both access to the numbers of identification discs necessary and the depth of experience necessary to muster the courage to take on such a project solo.

First and foremost in converting the dream to a reality has been my co-author Joseph W. Stahl. Not only did he have a deep knowledge of the topic backed by a significant collection of identification discs of his own, but he had a network of friends, acquaintances and fellow collectors whose assistance was essential. More importantly, he was easy to work with, willing to listen, patient, glacial to anger, and always receptive to a possible compromise. Without those traits it would have been impossible to collaborate on almost every page of a book. Thanks, Joe!

It is very difficult to decide which block in a structure is the most important, since the removal of any one may bring down the whole. Clearly though, Bob Marcus and Rich Burnham provided the foundation for this effort. When I ran into Bob, he had already moved on from an extensive collection of identification discs and had literally forgotten more about them than I knew. Fortunately, although he had distributed much of his collection, he kept extensive records which he more than generously shared. In three years of pestering, Bob has always provided the benefit of his expertise whenever asked for an opinion or advice — and incredibly without complaint.

Just as fortunately, Bob passed a healthy chunk of his collection on to Rich Burnham, who already had many discs of his own and who continued to collect even after he had accumulated what must be the largest collection of Union identification discs ever assembled. Even better, Rich has a digital camera, even more extensive records, and a generosity of heart that made both photographs and records available to us without reservation or a request for compensation. Rich's collection produced most of the images of the rarest discs included in this work and a plurality of the discs included within our inventory.

As if the foregoing was not generous enough, both reviewed portions of the text and provided guidance and necessary constructive criticism. Without the help of Bob and Rich this book may have remained a dream and certainly would have been a more meager effort. Both gentlemen have our sincerest gratitude.

Beyond those two, I have been blessed with a host of helpers. Nancy Dearing Rossbacher Sylvia, editor of the *North South Trader's Civil War* magazine, without any expectations, researched her publication's archives and sent me copies of many of the articles referenced in our text. George Fuld and Russell Rulau were open minded enough to discuss at some length a minor issue found in their essential *Medallic Portraits of Washington*, 2d ed., and in the process provided helpful guidance. Bruce S. Bazelon, another well-respected author, spent a considerable amount of time on the telephone discussing identification discs in general and the disposition of the Scovill button dies in particular.

The following people (in no particular order) also generously offered time, information and expertise: John Gibson, Rich Hartzog, Joseph Levine, Ron Coddington, Arthur W. Bergeron, Jr., Randy Hackenburg, Laura Katz Smith, Paul Cunningham, Dr. Lon Keim, Doug Watson and the kind staff people with the Numismatist, E-Sylum, and the United States Military History Institute.

My secretary, Brenda McDonald, has always been ready to drop what she was doing, run upstairs to my office, and explain how to do something on the computer, a device about which I am obstinately ignorant.

Most important of all, however, has been my wife, Joy Maier, who not only has provided tremendous assistance with the mysteries of word processing, but has trekked along to dozens of shows, helped with internet research, acted interested about the minutia of identification discs as I used her as a sounding board, smiled throughout, provided encouragement, and best of all didn't once complain about the conversion of retirement funds into identification discs.

—L. B. M.

*****

As Larry wrote, many people think about writing a book. Often it is the ones who have friends that push them to "write it down." For example, while I had access to numbers of identification discs, I had not studied or collected by types of discs, at which my co-author Larry B. Maier was an expert. Larry had spent a significant amount of time and study on the subject which he brought to the task. More importantly, he was willing to work by phone, e-mail and postal mail, plus meet at shows and in-between places to organize, discuss, comment on and argue about photos, text, and structure. All of this in a civilized manner which is very necessary and helpful as two adults try to produce something they both are very interested in and have opinions about based on past experiences. Thanks, Larry!

To the many individuals who read and liked my articles about soldiers and their identification discs and said to "keep telling the soldiers' stories," To Bob Moore, who first told me to write about what I was researching and to pass on the knowledge. To Bill Turner, who pushed me to collect and learn. To Joe Fulgni, who sold me my first identification disc. To Rich Burnham, who encouraged my first efforts at writing about discs. He was the person in whom I discovered a kindred collector and dealer with a collection to be envied. Others who encouraged me to write a book, in no particular order, were John Craig, John Hiller, Bradley Gernand, Bob Marcus, Lewis Leigh Jr., Randy Hackenburg, Ron Wolford, D.P. Newton of the White Oak Civil War Museum and Paul Bradock. And finally, to the Aiken and Nusbaum families, who on numerous holidays and get-togethers listened to my rambling and told me to "write it down and share the knowledge," this book is the result of your prodding and encouragement.

—J.W.S.

*****

Finally, we both would like to express our sincerest gratitude to Edwin C. Bearss for generously taking the time to review our manuscript, prepare a foreword, and provide helpful advice, and for his outstanding contributions towards preserving and enhancing interest in, and knowledge about, our nation's Civil War heritage.

TABLE OF CONTENTS

Acknowledgments	vii
Foreword by Edwin C. Bearss	1
Preface	3
1 • Introduction to Identification Discs	5
2 • Descriptions and Images of Identification Disc Styles	24
3 • Soldiers and Their Identification Discs	76
4 • Authentication	149
5 • Survey of Identification Discs by State and Style	165
Epilogue	189
Chapter Notes	197
Bibliography	205
Index	207

"The dead were then unrecognizable [following the Battle of the Crater on July 30, 1864] except by **medals** or letters found upon them. Men were torn into all shapes, and as black as could be, swollen out of all proportion, covered with flies and maggots, and emitting a stench sickening in the extreme. We found Augustus T. Thornton, of our company, but his body was too rotten to bring inside our lines, and we were obliged to leave him to be buried with the rest. I removed what few trinkets he had about him, and his **medal**, and a few days after sent them home to his Father. He was a good soldier, and though laboring under the disadvantage of being a little deaf, never shirked his duty, and had at last given his all in defense of his country."

— Corporal George H. Allen, 4th Rhode Island Volunteer Infantry
Forty-Six Months with the Fourth R.I. Volunteers in the War of 1861 to 1865
(Providence, RI: J.A.&R.A. Reid, Printers, 1887;
reprinted Higginson, 1998)

FOREWORD

by Edwin C. Bearss

One of the most popular items associated with the Union soldier treasured by collectors, Civil War aficionados and antiquarians is the identification disc. Known to millions of GIs and sailors of the 20th century as dog tags, they have been worn around the necks of Americans, as well as other nation's fighting men and women, since World War I.

As a World War II Marine, one of the first items issued to me upon my April 1942 arrival at the San Diego Marine Recruit Depot were two small circular dog tags. Stamped into the metal were my name, blood type, service serial number 374371, and religious preference. These two in number were worn suspended around your neck from a small metal chain secured by a clasp. When overseas in a combat zone, the necklace chain was usually replaced by a shoelace and the tags tied separately, thus abating any noise. It was understood that if you were killed one tag would accompany the body and be secured by the graves registration people to the temporary grave marker be it a cross or Star of David.

When I was wounded on Cape Gloucester, New Britain Island, my dog tags accompanied me though the next 26 months as I passed though a series of military hospitals. Prior to the 1st Marine Division's Cape Gloucester landing, as the unit was assigned to the Sixth Army, our circular naval dog tags were replaced by rectangular army identification discs with the usual basic information stamped into the medal that seemingly had a high content of tin and aluminum.

These, my last set of dog tags, accompanied me as a preserved reminder of the most memorable days of my life until lost following my mother's death in March 1962. What made them especially significant was my memory of our January 2, 1944, fight for Suicide Creek. Until their loss, the tags and shoe lace were blood-stained.

Needless to say the story of my dog tags and their fate whetted a latent personal interest in Civil War identification discs. Why latent? When I first became a Civil War fan, I read widely on the subject beginning in the winter of 1935-36. Then professionally, even more so, when I began a more than 40-year National Park Service career as a historian on September 28, 1955.

Intense involvement with the material culture of the Civil War soldiers and sailors began with my years at the Vicksburg National Military Park and the rediscovery and raising of the Union ironclad *Cairo* with its treasure trove of thousands of artifacts. About the same time I married Margie, whose interest, fascination and knowledge of the material culture of that war's military far exceeded mine. She took the lead in curating and cataloguing the thousands of artifacts that had belonged to the *Cairo* and her crew. Thanks to her I was introduced to the significance of the Union War identification disc.

My recognition of the discs' importance to battlefield archeology was reinforced by visits to White Oak Civil War Museum at White Oak, Virginia. D.P. Newton and his father, in their unsurpassed research and investigation of the extensive winter of 1862-63 encampments of the more than 120,000 men of the Army of Potomac, had recovered and documented a surprising number of discs.

My interest in Civil War identification discs and appreciation of their potential to contribute a new dimension to enhancing our knowledge about a particular soldier became apparent in the autumn of 2002. Since its inaugural issue in July 1989 I have served as an editor for the *Gettysburg Magazine,* a popular publication featuring articles highlighting America's best known Civil War campaign and battlefield. The publisher had sent me for review and comment a manuscript by Joseph Stahl. It fired my interest as Stahl demonstrated the importance of identification discs as a research tool to get to know the soldier who once wore the tag. Impressed by what I read, I strongly recommended Stahl's submission for publication, and it appeared in the January 2004 issue and was titled "Pvt. Theodore Sanborn, Company D 12th New Hampshire Infantry." The popularity of the subject with the readers was such that the publisher determined to feature in future issues of Joe Stahl's "dog tag" series.

I was enthused to learn from Stahl that he and Larry B. Maier had employed their research methodology and writing skills, coupled with networking with other collectors, to co-author *Identification Discs of Union Soldiers in the Civil War.* I could think of no two better people qualified to accept this challenge. I was honored when Joe Stahl invited me to prepare the Foreword. On reading their draft I was not disappointed. *Identification Discs* will be must reading, as an invaluable reference book for collectors, curators, and antiquarians. More important, it will be a ready source for those Civil War enthusiasts and students as well as the general public who are fascinated by the material culture of the Union soldier and the story the items can tell about the long-dead person to whom they once belonged. The chapter titled "Soldiers and their Identification Discs" will prove invaluable to those who desire to hone their research skills in following a soldier's or ancestor's Civil War career.

Edwin C. Bearss is chief historian emeritus for the National Park Service.

Preface

The intent of this book is to provide a comprehensive study of one of the most collectible and prized keepsakes of the American Civil War — the identification discs that were worn or carried by many Union soldiers. Although other relic books have frequently shown a few examples of identification discs, this book is the first to concentrate solely on the subject of Civil War identification discs.

These identification discs provide us with tangible reminders of these soldiers and their service more than 100 years ago. The introduction discusses the necessity for, and development of, identification discs; how and by whom they were made and marketed; and how they were worn by the soldiers who purchased them. In addition, comparisons of information contained on certain discs against the records of soldiers known to have worn particular substyles have permitted conclusions to be drawn as to when those substyles first became available to the troops. As expected, some types were sold very early in the war, while some became available much later.

Identification discs came in many styles and substyles. The authors have for the first time attempted to catalog all known examples in a systematic fashion and in a manner that will offer the collecting community common terms for ease of reference and discussion. Photographs of both sides of each substyle are provided so the historian, curator, and collector may compare those depicted for authentication of any disc in his possession or those under consideration for purchase. Also included is a relative measure of the rarity of each substyle as compared to the collection of discs referenced in the text.

This volume includes brief service histories for more than 100 soldiers who are known to have owned identification discs, as well as photographs of each soldier's disc and in some instances, images of the soldier.

There is a chapter on collecting identification discs. It offers information that will be helpful when one is attempting to evaluate the authenticity or condition of a disc being offered for sale or trade. Also reviewed are the factors that may impact the value of a particular disc. This information will help the reader, with practice, to identify and avoid fakes, frauds and reproductions made as souvenirs for modern re-enactors.

More than 600 identification discs were reviewed in the preparation of this work. Tables document the results, with one organized by state and regiment and the other by style and substyle. The first table, which is sorted by the unit's state, will allow quick confirmation of other known discs from a particular unit, while the second table, sorted by the style and substyle of identification disc, will provide the reader with a comparison with any disc offered to other known discs of the same substyle. With each disc is other basic information such as the owner's unit, company and type of disc. The tables cover known discs from more than 250 infantry regiments, cavalry regiments or artillery units. Also included is each company in which at least one soldier of that unit bought an identification disc, while some units have known discs from several companies.

The Epilogue contains the stories of four soldiers, each of whom owned a disc included and depicted within the text. Each of these soldiers served in a different regiment that comprised one brigade that fought over "Burnsides Bridge" on September 17, 1862, during the battle of Antietam. Where they served and what happened to each one is chronicled. Each individual story presented is typical of the history that resides in these small pieces of metal.

Introduction to Identification Discs

Most of the 30,000 untested Federal troops on the march from Washington, D.C., to Centerville, Virginia, acted as if they thought they were on a lark. As they tramped along, dressed in varied, colorful, and sweltering uniforms, only the most pessimistic dwelled on fate or mortality. So confident were the Yankees that some discarded all their equipment but muskets, bayonets, canteens and cartridge boxes, viewing the rest as superfluous for a campaign that should take less than a week, or maybe two at the most. The procession moved so slowly that many had enough time to drop out of line to fill canteens and a few to pick berries along the Warrenton Turnpike and still get back into line before their sergeants noticed they were missing.

It took two and a half days, from July 16, 1861, to July 18, 1861, for the Union army under Brigadier General Irvin McDowell to cover the same twenty-five miles that would become a routine day's march by the end of the war. The Union command wasted another two days reconnoitering the area around the Bull Run Creek while stragglers from the column trickled into camp. As the Yankees sat around guttering camp fires the night before their first battle, there was plenty of nervous bantering about courage, scadaddling and "seeing the elephant" and not a little grumbling by those who had consumed their week's worth of rations while still on the march.

Union General McDowell and the Confederate commanders, Brigadier Generals Pierre Gustave Toutant Beauregard and Joseph E. Johnston, all had basically the same plan for the morning of July 21, 1861. Each side planned to attack their adversary's left flank, and as a result each side strengthened its own right. Had both plans succeeded, the two armies might have spun around the stone bridge at the center of the field like a pin-wheel.

The Union army struck first, driving the thin Rebel left flank back towards its center in a withdrawal that teetered on the verge of a collapse. The Federals were only denied their anticipated quick and easy victory when a Southern brigade, under the command of Brigadier General Thomas J. Jackson, held back the Northern advance like the figurative stone wall that would soon thereafter become Jackson's new, unofficial, first name. For several hours the Federals marched up Henry Hill by regiments. But each was shattered in its turn by volleys from the determined Confederates.

While the Northerners made one futile attack after another, Beauregard and Johnston changed plans and pushed reinforcements beyond the Yankee right flank. When the Rebels cracked back on the exhausted and shot-up Union line, it quickly crumbled, sending thousands of panic stricken Federals running for the rear.[1] Most covered the same twenty-five miles back to Washington by early morning on July 22nd.

Psychologically the retreat was undoubtedly far longer than the hours it took. While the men pushed one leaden foot ahead of the other, those who weren't too exhausted to think tried

A modern image of the Stone Bridge at 1st Manassas taken from the southwest.

to process the day's experiences. As the rain that began to fall after midnight soaked through their coats, and rivulets of cold rain dribbled down their backs, they pondered their humiliation, their mortality, and comrades who lay strewn over the battlefield. Almost as bitter as the knowledge of their death, friends and in some cases relatives, had been left in the hands of the enemy to suffer the ignominy of a nameless burial in mass graves. To die for one's country was one thing — to risk that sacrifice was an obligation and a badge of honor. To anonymously molder in a shallow grave without mourners, flowers or loved one's tears was another matter altogether.

The full extent of the disaster was difficult for most Northerners to comprehend. Four hundred seventy Bluecoats were known to have been killed in action and abandoned on the field, while many hundreds more were left behind as missing or wounded. Both soldiers and the public were shocked. For perspective, the number of killed in action on that lone July day was roughly ten percent of all the 4,435 Americans killed in action during the Revolutionary War and nearly thirty percent of the 1,548 who didn't return from the Mexican War because of wounds.[2]

Most of those 460 Union corpses were dumped into anonymous graves because at the time the United States government did not issue any form of personal identification.[3]

While Major General George B. McClellan spent the balance of 1861 reorganizing and retraining the Eastern Army of the Potomac, the War Department slowly began to consider the issue of the identification of remains. Its first attempt was rather feeble. General Order No. 33, issued on April 3, 1862, directed field commanders to set aside burial grounds in close time and proximity to any battle, and "'where practical'" to place some sort of headboard at each grave with the dead soldier's name written or inscribed.[4]

Soon after the battle of 1st Manassas, some of the Bluecoats began taking steps to remedy the identification problem themselves rather than rely on their government. Various methods and objects were pressed into service. Many men purchased brass stencils which they used to ink their names, and often their regiments, on items of clothing and equipment. Stenciling provided excellent theft protection while a soldier lived and identification if fate turned her back.

Others fashioned home-made identification badges and discs from coins or even by carving their information on small pieces of bone. Some purchased or were given Bibles and books on military tactics. Often the Bibles came with a blessing that included the soldier's name and regiment, along with expressions of encouragement, love or faith penned by the donor. Those soldiers not so blessed, or who purchased their own books, often added personal information on the assumption that if one was going to carry a book into battle anyway, it might as well also provide identification. A few charities even inserted a preprinted message with appropriate blank spaces inside the cover which prompted and facilitated the recipient to insert pertinent information should he fall in battle.[5] Others groups actually distributed preprinted paper tags, similar to coroner's toe tags, with spaces for information and wire for easy attachment.[6]

Small tintype image of an unknown soldier wearing an identification disc.

American enterprise and ingenuity also quickly responded to the need. Flags had hardly been changed over Fort Sumter when Northern newspapers and weekly magazines began to include advertisements for identification badges. Those early badges were generally small, silver, shield shaped pins which could be purchased by someone at home, inscribed by a jeweler with the soldier's name, company and regiment, and then given or mailed to the soldier.[7] Later,

Left: Private Norris Myers' brass identification stencil. *Right:* Lieutenant Yeaman Jones' carved bone identification badge.

A Bible presented to Henry Myers with an identification label.

identifying information would be engraved upon one of the geometric shapes assigned to a soldier's corps.

Identification badges were a good idea, but had their drawbacks. They were generally not available in the field where both engraving tools and the skills to use them were equally rare. Perhaps just as important, they were relatively expensive, which put them out of reach of most soldiers who struggled to provide both for themselves and their families on a private's monthly pay,

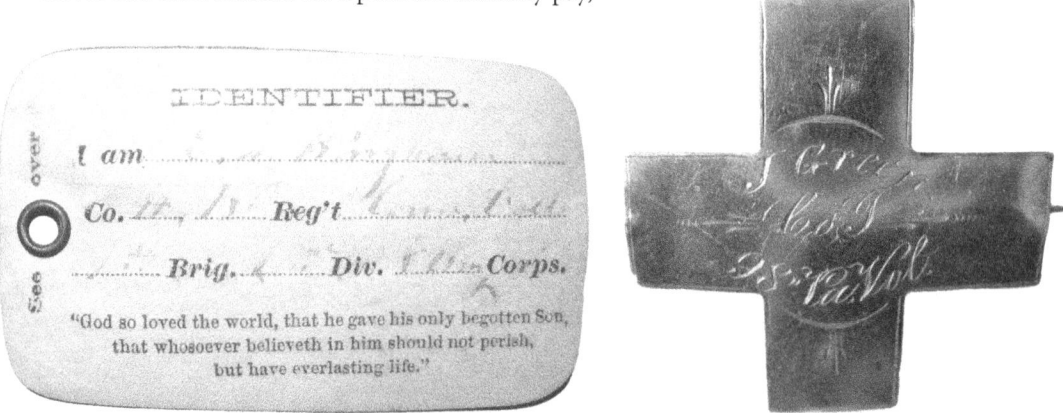

Left: Private Bingham's paper identification tag (courtesy Lewis & Rosalie Leigh, Jr., Collection). *Right:* Private Jacob Green's 6th Corps identification badge.

which, when it arrived, was usually months late. Based upon those that survived, it seems the majority of identification badges were purchased for, or by, officers, who tended to come from a higher socioeconomic background.

At some point very soon, or perhaps even slightly before, the First Battle of Bull Run (1st Manassas), some clever merchants saw the potential demand for a cheap, light and durable way to provide identification that did not require the skill of an engraver, or the wallet of an officer. The identification disc, the grandfather of the modern military dog tag, quickly began to appear.

Although there is no direct evidence, it can be surmised that the very first identification discs were small, half dollar shaped brass coin-like medallions that bore the likeness of Major General Winfield Scott on the front (obverse) with a hole drilled above his head for suspension. On the other side (reverse) a blank surface was provided for the placement of identification information. It appears that these discs were marketed directly to the men by their regimental sutler or by a merchant set up in or just outside their training camps. (See Chapter 2.) It is believed that the Winfield Scott (12A) style was probably the first because Scott was the general-in-chief of the army from the opening shots at Fort Sumter until Major General George B. McClellan assumed field command on July 27, 1861, and total control over the army after Scott resigned from the army on November 1, 1861.[8] Despite the fact that Scott was still considered a hero from the Mexican War, it is doubtful there would have been any significant market for discs bearing his likeness after November 1861.

Sutlers were itinerant merchants who were granted permission, usually by its colonel, or later by the particular brigade commander, to travel with a particular regiment. For that privilege, and the opportunity to charge profiteer prices, the sutlers were required to pay a fee out of their profits which was supposed to be used to acquire supplies for their regiment's field hospital.[9]

By the fall of 1861, of 200 regiments inspected by the Sanitary Commission, 188 had a sutler, of which 103 were appointed by the colonel of the regiment.[10] Almost certainly, Union soldiers began to purchase identification discs from their regimental sutlers in the winter encampments of 1861–1862.

The first discovered written account of the purchase of an identification disc from a sutler was found in a letter written by Henry L. Franklin of the 2nd Vermont Infantry which he sent home in February 1862. Franklin purchased two discs, each bearing the image of a Displayed American eagle on the obverse and his name, company, regiment and possibly his home town stamped on its blank reverse.[11] The Displayed Eagle style (5A), with wing tips up and a Union shield on its chest, was the same basic image that appeared on the reverse of United States $2.50, $5.00 and $10.00 gold coins at that time.

About his new acquisitions, Franklin provided this explanation to his mother: "'You will find a medal that are made for soldiers to wear around their neck in time of battle and if they are killed you will know who and where they belong, I have 2 and I send you one of them.'"[12]

Not all the soldiers who bought identification discs hung them around their necks as Franklin described. Along with their discs, some purchased a suspension pin either in the shape of an American Eagle or in the shape of a shield bearing the likeness of either Brigadier General Philip Kearny, McClellan or later of Major General Joseph Hooker, from which the disc could be suspended and pinned on their chests.[13] It is easy to conclude that many of the discs now being recovered by relic hunters with metal detectors were lost when pins popped open, broke, or got snagged on bushes or tree branches. In fact, the suspension pins are much rarer than the discs themselves, and the large number of those offered for sale were dug.

Even before Henry Franklin bought his disc, it appears that sutlers were offering other styles in addition to the Eagle 5A style. Most were made from brass like the Scott 12A but a few

A drawing of a sutler shack by Edwin Forbes.

were also minted in white metal.[14] For appearance, the former were usually gilded in gold while the latter were coated with silver wash. Those new styles included (in no particular order) a disc with the Union shield and the date 1861, several with the likeness of George Washington and a multitude of slightly different designs revolving around the bust of General McClellan.

Left: The reverse side of an 1861 U.S. $10 gold coin. *Right:* An obverse view of an Eagle 5A identification disc.

Several factors support the conclusion that the Eagle (5A) and the Washington (3A) discs were, after the Scott design, the first to be made available to Union soldiers. In addition to the above referenced letter, military records for soldiers whose discs have been investigated support that observation regarding the Washington 3A disc. Private Daniel Cummings of the 9th New

Left: An unknown private wearing an identification disc on his watch fob chain (note what may be a Company Letter H above the second badge) (courtesy U.S. Army Military History Institute). *Right:* A tintype of a soldier with his identification disc suspended from his button hole by a string.

Figure 8: A soldier with his identification disc suspended from a ribbon.

Left: Private Paine's identification disc suspended from an Eagle Pin (courtesy Lewis & Rosalie Leigh, Jr., Collection). *Middle:* Private Fowler's identification disc suspended from a McClellan style suspension pin (courtesy Lewis & Rosalie Leigh, Jr., Collection). *Right:* Lieutenant Zeisert's identification disc and Kearney suspension pin (courtesy Brent Musser, Jr., Collection).

York Volunteer Infantry presumably purchased his Washington (3A) disc sometime between his enlistment on April 23, 1861, and his disability discharge on August 20, 1861, perhaps only days or a few weeks after 1st Manassas. A boy from Lancaster, Pennsylvania, called Peter H. Flick somehow ended up in the 70th New York Volunteer Infantry where he purchased a Washington (3A) disc. Unfortunately, Private Flick was denied the opportunity to enjoy it for very long. He enlisted on July 10, 1861, but drowned in the Potomac River on March 17, 1862.

Hooker, McClellan and Kearney style suspension pins.

Soon after the appearance of the Eagle (5A) and Washington (3A) discs (or perhaps contemporaneously), Union soldiers were also given the opportunity to choose as an alternative the Shield 1861 (2A) and the McClellan (1A) styles. Presumably, the Shield 1861 was first marketed in that year because it was either replaced or joined by a similar design bearing an 1862 date (Shield 2B). It is doubtful the McClellan styles appeared before "Little Mac" assumed field command of the army in July of 1861 or overall command on November 1, 1861. He remained commander of the Army of the Potomac until President Lincoln replaced him on November 7, 1862, because of McClel-

Left: An unknown private wearing his identification disc below his 6th Corps Badge. *Right:* Private H. Phillips of Company H of 9th Iowa Infantry wearing an identification disc (both courtesy U.S. Army Military History Institute).

lan's timidity and unjustified arrogance. Although some of the Union soldiers continued to idolize McClellan throughout the war, his marketing appeal certainly diminished after he was sacked.[15] It may also be presumed that since those four basic styles, excluding the Scott, account for the majority of all surviving identification discs, they represent the largest numbers sold and therefore they were available for purchase for the longest period.

Military service records for some of the soldiers who are known to have purchased identification discs while in the service also provide information which can help estimate when certain style discs began to appear in sutlers' shops.

Mortimer Doud of the 151st Pennsylvania Volunteer Infantry presumably had his McClellan (1A) disc stamped before he succumbed to disease on December 29, 1862. First Sergeant John W. Jenkins of the 48th P.V.I. bought his McClellan (1B) disc before he was discharged on December 4, 1862, after a shell fragment fractured his skull at Antietam.[16]

According to his military records, Lorenzo D. Finch, of the 72nd New York Infantry regiment, went to his sutler's shanty, wagon or tent to get a Washington (3A) style disc stamped with his vital information before December 1862, the date he was transferred to the 4th Independent New York Battery. Presumably he would not have given instructions for the sutler to stamp the number of his former unit on the reverse.[17]

James McInnes of the 5th Ohio Volunteer Infantry purchased his Washington (3C) disc before December 22, 1862, the date on which he was discharged for chronic diarrhea.[18]

Even though it was relatively rare for an officer to purchase an identification disc rather than a badge, Lieutenant John Mitchell probably obtained his Washington (3B) disc sometime in late 1861. On May 3, 1861, he mustered into Company A of the 71st New York State Militia, a 90-day unit from which he was discharged on July 30, 1861. Then, on November 1, 1861, he mustered into the 101st N.Y.V.I. as a lieutenant in Company D. Apparently he only had the sutler stamp his name and Company A because he knew he was leaving the 71st but didn't know to which regiment he was headed. Curiously, he mustered out of the 71st as a private and into the 101st as a lieutenant. Admittedly, it seems odd that he would have received his commission before assignment to his new regiment. But whatever the explanation for not having his regiment stamped on his disc, he had it before September 1, 1862, when he received a medical discharge for rheumatism contracted at the battle of Savage Station.[19]

Several months before some of the Yankees listed above purchased their discs, the United States War Department was provided with at least one opportunity to authorize a standardized identification disc. On May 3, 1862, on the eve of the bloody Peninsular Campaign, John Kennedy of New York City wrote the following letter to War Department Secretary Edwin M. Stanton:

> Enclosed please find a plan of a badge or medal which I propose getting up for distribution in the Army for the purpose of obviating the difficulty already experienced at Bull Run in identifying the bodies of the slain after buriel [sic]—[the medal] Is intended to be of metal and worn by the soldier under his clothing. It will simplify the record kept by the buriel [sic] squad and enable the friends of the Patriotic dead to identify their remains even in after years.... I have no doubt sir but your intelligence will enable you to see at once that their [sic] can be no difficulty hereafter in identifying [sic] the bodies of soldiers wearing the "Kennedy badge." The object of this letter is to solicit permission from the department to visit the army with the necessary tools to strike them off and distribute them to the officers and men.... As the object to be attained suggests under the existing state of affairs promptness of action I would most respectfully request and [sic] early reply and a favorable consideration.[20]

A small drawing accompanied Kennedy's solicitation. The identification disc he envisioned bore a striking resemblance to the extremely rare Double Blank (16A) disc worn by a few Union Soldiers. Three days after it was sent, Kennedy's suggestion was bluntly rejected by one of Stanton's assistant secretaries without explanation. Kennedy's request was probably declined because the need was already being met by free enterprise and/or, because the War Department was skeptical of such a proposal made by a man whose spelling was so bad.[21] Unless the Double Blank style disc was his, it does not appear that Kennedy pursued, or made any money, from his plan.

Based upon what was to follow, one can justifiably argue that the U.S. government made a serious mistake by rejecting the concept, if not Kennedy's specific proposal. Before the close of 1862, the Federals would suffer a number of major defeats where the remains of Union soldiers were abandoned and left on the field of battle. At the end of August at 2nd Manassas, the Yankees lost 1,747 killed and 4,263 missing in action. Three and a half months later, 1,284 men were left dead on the slopes in front of Marye's Heights at Fredericksburg and another 1,769 were reported as missing.

Unknown soldier from Company E of 11th Iowa Infantry wearing two identification discs (courtesy U.S. Army Military History Institute).

In 1863, the Union suffered two major defeats where the Federals were denied the opportunity to inter their fallen comrades. At Chancellorsville 1,606 Union men died and an astonishing 5,919 were declared missing in action. In another battle also fought predominantly in heavy growth, 1,656 men were killed at Chickamauga while 4,774 of their comrades were never officially accounted for.[22]

The carnage grew in 1864. At the Wilderness, the remains of the 2,246 dead and 3,383 missing were abandoned to the enemy. Over the next week, 2,725 men were killed and another 2,250 were reported missing around the Spotsylvania County Court House.[23]

At Cold Harbor, the last major battle where the Federals surrendered possession of the field to their foes, but only after ten days, the statistics show 1,844 dead but only 1,816 men missing in action.[24]

Long after those grim statistics had been accumulated, the last styles of discs offered for sale appear to have been minted in late 1864 or early 1865. Of the three "No Compromise with Armed Rebels" (Union 8A) style discs observed by the authors, two were purchased by very late enlistees. William Graham of the 56th N.Y.V.I. didn't enlist until March 29, 1865. Even later than that, Jacob Boshart must have bought his soon after enlisting, with exquisite timing, into the 80th N.Y.V.I. on April 10, 1865.[25]

Military records also indicate that as late as August 31, 1864, when Benjamin Cooper enlisted in the 1st Rhode Island Light Artillery, at least his sutler was still selling the same Battle disc with hanger tab (Battles 4C) style that George Bickel of the 9th Indiana Volunteer Infantry purchased before he was discharged for illness on October 11, 1862. Even substitutes were concerned about their remains being identified, as shown by Private Neal McDevett, who purchased his Battles (4B) disc sometime after he mustered in on September 27, 1864.[26]

Other styles of discs practically dated themselves. Two McClellan discs came with the date on which the battle at Antietam Creek was fought, (McClellan 1F & 1G), thus dating their manufacture after September 17, 1862, but probably not many months thereafter. A different style McClellan, obviously done by a different die-sinker, first bore the legend "The War of 1861, & '62." The next year the design was freshened by the addition of the year 1863. Apparently not wanting to scrap a perfectly good die hub, this style was amended one more time for 1864 (McClellan 1I, 1J, and 1K). The Displayed Eagle style with "Union" between the wings (Eagle 5B, 5C & 5D) followed the same pattern by adding a four to the prior years "1861, 2 & 3." Finally one of the Battles (4A) discs included the minted notation of battles fought in "1861, 2 & 3" while one of the Grant discs included the year 1864 on its obverse (9C).

As the war progressed, other disc styles were minted to replace or supplement those listed above in hopes of appealing to the changing tastes of potential consumers. After the men had a few battles under their belts, the sutlers offered discs specifically configured so that a list of their actions could be added (Battles [4A-D], McClellan [1C], Shield [2C]). Representations of President Abraham Lincoln (6A), General Franz Sigel (10A), Major General Nathaniel Banks (7A), Major General Joseph Hooker (11A) and Lieutenant General Ulysses S. Grant (9A-C) all appeared on different discs, presumably purchased by men who served under them, who developed new heroes, or who didn't want the image of a disgraced general around their necks or pinned on their chests. But just to be sure no one would be offended by the image on the obverse, the sutlers even added additional patriotic designs with shields, eagles, state seals, and even sold discs that were manufactured with both sides blank.

How the sutlers obtained their inventory of discs, what they charged, and the details of how they stamped them, still remains the subject of conjecture. Only two advertisements for true identification discs have been located. In his definitive text on corps badges, Stanley S. Phillips provided copies of two advertisements which, along with other style badges, displayed pictures of the same type disc with a hanger tab bearing the bust of Ulysses S. Grant surrounded by the words "Unconditional Surrender."[27] Phillips didn't provide either the name of the publication in which the advertisements appeared, or the dates they ran. A rough estimate of when this disc, referred to in the advertisement as a "soldier's charm," first became available can nevertheless be deduced. Grant didn't become known as "Unconditional Surrender Grant" until after the capture of Confederate Fort Donelson in mid–February 1862, when he demanded, and eventually received, "No terms except unconditional and immediate surrender."[28] One of the two advertisements also contained images of corps badges, which were not introduced until the spring of 1863 by Major General Joe Hooker, while the other referenced the Presidential election of 1864.[29]

No advertisements for any of the other forty-eight substyles of identification discs have definitively been located. Several advertisements that appeared in the Harper's Weekly Magazine may have been for identification discs, but since no pictures were included, the issue remains unresolved. On May 25, 1861, W.A. Hayward requested orders for "Union shields, eagles, etc." Then on August 10, 1861, he proclaimed that he "Wanted 1000 agents to sell miniature pins of Gen. Scott, Butler and all the Heroes."[30] Four months later, Henry Osborne, of 58 Nassau St. N.Y., N.Y. solicited for agents to hawk his "'New McClellan Medal.'"[31] During the same period of time, partners B. & H.D. Howard offered "Sutlers Supplies ... armorer's oil, soldiers companions, gloves, shirts ... tobacco, pistols ... and a full assortment of goods for Sutlers."[32] We will probably never know for sure whether their inventory included identification discs.

At least one enterprising Yan-

Above: A soldier with his identification disc suspended from an eagle pin. *Right:* A cavalry soldier with his identification disc suspended from an eagle pin.

Henry Blanchard Letter, dated July 29, 1862, ordering identification discs from his father (courtesy Lewis & Rosalie Leigh, Jr., Collection).

 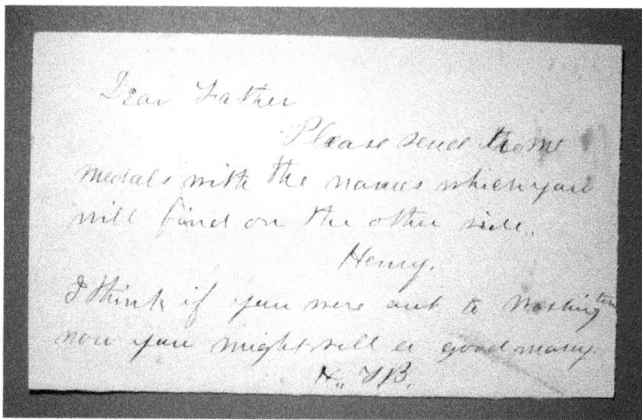

Left: The identification disc made for Henry Blanchard by his father. *Right:* A brief note Henry Blanchard wrote to his father about identification discs for his unit (both courtesy Lewis & Rosalie Leigh, Jr., Collection).

kee, Henry T. Blanchard, of Company K, 2nd Rhode Island Volunteer Infantry, decided that he had the right to help his buddies solve their identification problem while at the same time keeping the money out of the hands of the generally despised sutler. Blanchard worked out an arrangement with his father where the latter would fabricate small oval copper discs and stamp on the vital information obtained from the soldier upon orders supplied by his son Henry. After the home-made discs arrived in the mail, Henry would then deliver them to the purchaser and hopefully collect the money.

The Blanchards' business must have been fairly successful, based upon inferences which may be drawn from a portion of Henry's July 29, 1862, letter to his father. "I have got a few names for medals which if you please you can send out. Of the rest which you already sent out I have got the pay for all but two, except the 6 which were lettered wrong. I have them yet but suppose I might as well throw them away as to keep them. The names I have got now are as follows, Ordly. Sergt. Obed H. Gifford Co. K. 2nd R.I. Vol enlisted June 6th 1861, James F. Thurston Co. K. 2nd R.I. Vol. Enlisted Sept 10th, 1861, Corporal John J. Hilton Co. K. 2nd R.I. Vol. enlisted June 6th 1861 ...p.s. two for Hilton, Percy Millar Co. C. 2nd R.I. Vol. Enlisted June 6th, 1861 ... Corporal L. C. Belding Co. K. 2nd R.I. Vol enlisted June 6, 1861."[33] Unfortunately, Henry Blanchard didn't include the price he charged in the letter to his father.

It does not appear that those sutlers without competition in the ranks obtained their discs from advertisements by mail or Adams Express. The most likely other possibilities were that either wholesalers had them at the depots where the sutlers obtained their wares or that traveling agent/salesmen brought them to the sutlers in camp. It is unlikely, though, that most of the sutlers got their discs from a central wholesaler because there was a diversity of styles within corps, divisions, brigades, and even among soldiers in the same regiment. If all the sutlers bought from only a few suppliers, presumably most of the regiments would have had similar styles for a given period of time.

The sutlers' most common source for their blank discs was probably through traveling salesmen, or agents as they were called in many of the badge advertisements, who brought the discs to the sutlers, probably during winter encampments when the sutlers remained stationary for months. It may be safely presumed that the diversity of disc styles occurred either because some sutlers decided to have a variety on sale at any one time, or because events dictated new styles more appealing to the soldiers.

What the sutlers paid the salesman for their discs can only be projected based upon the advertisements noted. Both contained solicitations for agents, and both listed the price for an unengraved Grant disc as $1.00.[34] Presumably that was the wholesale price. Francis Lord, in his work on sutlers, discovered a ledger with entries from August 9, 1863 until January 13, 1865, that he believed had belonged to the sutler for the 10th Massachusetts Infantry. Among a host of sutler's goods such as canned foods, cigars, scissors, ink and playing cards appeared these three entries: "1 badge—$2.50 ... 1 U.S. badge—$2.00 ... 1 Custer Badge—$27.00"[35] A price of $2.00 to $2.50 would have been consistent with the value added by stamping and a reasonable mark-up. It is also consistent with the wholesale prices charged for silver plated identification/corps badges.[36]

Compared to a private's pay of $13.00 to $16.00 per month, one can get a sense of how important to the soldiers the purchase of an identification disc really was (as well as the exorbitant prices charged by at least one sutler). For the same money a soldier could instead have gotten himself any one of; a bottle of whiskey or rye, a deck of cards, a pair of candlesticks, a pair of spurs or a knife.[37]

Because most sutlers were merchants rather than skilled craftsmen or jewelers, the manufacturers and distributors had to overcome a major problem, i.e., how was the soldier's information to be placed on the disc. It appears that this problem was solved by selling the sutlers a letter and number die punch kit along with a supply of discs. Although the authors have yet to observe a punch set or kit proven by unimpeachable provenance to have been from a Civil War sutler, or even to have spoken to any old time collector or expert who claimed to have seen such a kit, it seems highly logical that such sets or kits existed.[38]

A close study of the letter designs for similar substyles of discs from the same regiment, and also across different regiments, reveals subtle similarities among the fonts of certain letters used on similar substyle discs, indicating that different sutlers had and used the same punch kits for the same discs. The only logical way this could have happened was that the punch kit and the blank discs came together. It must be noted, however, that a few strays have been found which indicate that some sutlers may have continued to use a kit from a previous style with new blanks, perhaps because the blanks and kits were sold separately and his old punch kit was still functional.

Further, there are defects, most likely caused by weaknesses in the design or metal used to manufacture certain letter punches, that appear in different regiments but on the same style disc, indicating the same punch kit manufacturer. Finally, it appears that some kits came with complete phrases or battle names along with the usual individual letters and numbers. For example, the white metal Washington (3A) discs often have the same phrase "War of 1861" or the same decoration "-< >-" punched on the reverse with the soldier's other information. Those embellishments appear on most Washington (3A) discs across multiple regiments and states, but not on any other style, making it easy to surmise that each was done with a nearly identical punch that came strictly with that style disc.

History also tends to repeat itself. In the years just prior to 1917, the United States government decided to supply standardized identification discs to all our Doughboys. A sergeant from each regiment was issued a small wooden box which contained individual punches for all the letters in the alphabet, numbers 0 through 9, a small hammer, an anvil designed to receive and hold standardized aluminum discs and a small diagram showing how the information was to be arranged.[39] (See photograph on page 154.) Given that fifty years passed between the Civil and the War to End War, it can be assumed that no better technology to stamp discs was available during the Civil War.

Nevertheless, there is still some question about whether all Civil War sutlers stamped individual letters to form names and other information, as was done by First World War sergeants, or whether some stamped the discs with a machine or gang punch. In his multi-volume encyclopedia of Civil War relics, Francis A. Lord wrote that "lettering [on discs] was machine stamped."[40] The consistency and preciseness of letter placement, particularly the soldier's name, which generally was placed in a curve around the rim of the reverse, gives some credence to the possibility that the letters might have been placed in some sort of holder and then punched as a word or group of words.[41] It is also conceivable that some sort of frame, like a stencil, could have been provided to fit above the disc to guide the sutler when he punched. Such guides were provided with some of the World War I punch kits.[42] However, no such gang letter holder, gang punch or guide has ever been observed by the authors, nor has any sort of machine which would have been capable of punching words or a whole disc at one time.[43]

Also, because one can observe variations in the skill of placement of letters by different sutlers, it is doubtful that a holder or machine was used. Most likely, Lord used the term "machine stamped" to distinguish imprinting done with manufactured letter punches from jeweler's engraving done by hand on identification badges.[44]

Lord also provided an anecdote (perhaps apocryphal) which showed how adept sutlers became in stamping discs, either with or without machines. According to Lord, "[i]n July, 1864, the 14th New Hampshire Infantry, passing through Charlestown, West Virginia, on its way to the Shenandoah Valley, purchased many of these brass discs from a sutler who had set up his tent by the roadside. He stamped each disc purchased with name, company, and regiment."[45] Such a feat could only have been accomplished in acceptable time by well practiced hands.[46]

The issue of who minted the blank discs is only somewhat more settled. Even before the Civil War, the United States had a well established industry of medal and token production. Those products included hard times tokens, presidential campaign and commemorative medals and tokens made for advertising particular merchants or political opinions.

All of those products, and United States coins, were made in basically the same way. A highly skilled and usually talented craftsman, known as a die-sinker or die-cutter would create the die in a long process described by coin expert R.S. Yeomans. After a tracing of the design was made and transferred to wax

> [a] piece of steel is smoothed and coated with transfer wax, and the tracing impressed into the wax. The engraver then tools out the steel where the relief or raised effect is required. If the design is such that it can all be produced by cutting away steel, the die is hardened and ready for use. Some dies are not brought to a finished state, as some part of the design can perhaps be done better in relief. In that case, when all that can be accomplished to advantage and the die is completed, it is hardened, a soft-steel impression is taken from it, and the unfinished parts are then completed. This piece of steel is in turn hardened and, by a press, driven into another piece of soft-steel, thus making a die which, when hardened, is ready for the making of coins.[47]

Those "unfinished parts" were commonly letters, numbers, and symbols like stars, berries, bars or dots, which the die-sinker would stamp into the die with pre-manufactured punches because it was faster, easier, and more uniform to stamp rather than carve them. Either the die-sinker himself, or a larger manufacturer, would then place the completed working die in a minting machine and stamp out the completed coins, tokens or, eventually, identification discs.[48]

Fortunately, some of the die-sinkers included their names and/or initials on their creations. The most prolific name dropper was named Frederick B. Smith, who was born the day after Christmas in 1811. He kept his shop at 122½ Fulton St. New York, New York. From there he turned out several styles of McClellan discs (McClellan 1A and 1B), and one Lincoln disc (6A).[49] Each bore some variation of F. B. Smith tucked in small letters under the shoulder of the bust on the obverse.[50]

Another, and unusually talented, die-sinker was Joseph H. Merriam, who commenced his trade in Boston, Massachusetts, in 1850 at 18 Brattle Square at the corner of Elm. In addition

Left: A closeup of the die mark B. SMITH F. *Right:* A closeup of the die mark F.B. SMITH.

to the Washington 3A and Battles 4C identification discs, Merriam also produced letter and number punches, seal presses, brands, and stencils, and the Medal of Honor during his career.[51]

Rivaling Merriam in talent and the production of a handsome Washington disc was Robert Lovett, Jr., who practiced his art from 200 South 5th Street in Philadelphia, Pennsylvania. During the war he cut the Washington with Security below (3D) and Lady Liberty (15A) discs. He also produced the die for the Confederate Cent, a fact he tried to hide from his fellow Philadelphians.[52] In keeping with the family tradition, his brother, George H. Lovett, from his shop at 131 Fulton St., New York, New York, contributed the Washington (3E) disc.[53]

Unfortunately, none of the other die-sinkers chose to include their name or initials on their identification discs. This deficiency may be attributed to the common practice at the time of sharing master dies. Visually, it would seem that Merriam and Lovett shared their work with others who had the decency not to put their own name on a borrowed die from which they excluded the real artist's name.[54] Others apparently viewed their discs as products rather than art and therefore undeserving of recognition. For whatever the reasons, we are left to speculate who cut most of the Civil War identification discs sold and worn during the war.

One of the earliest identification disc styles, the Displayed Eagle with "United States" below and "War of 1861" above (Eagle 5A) could have had any number of fathers, but it seems most likely that it was inspired by the straight winged eagle that had appeared on the reverse of some American gold coins since 1807.[55]

Left: A closeup of the die mark SMITH B. *Right:* A closeup of the die mark SMITH F.

Left: A closeup of the die mark F.B. SMITH on a Lincoln (6A) substyle. *Right:* A closeup of the die mark MERRIAM on a Washington (3A) substyle.

The most likely candidate for die-sinker and/or manufacturer of at least the Displayed Eagle substyle disc (5A and possibly 5B, C, and D as well) was the Scovill Manufacturing Company of Waterbury, Connecticut.[56] That assumption is supported by two 29mm (the same basic size as most identification discs) advertising coins manufactured and marketed by that company; one for "Scovill's Daguerreotype Materials" firm (relationship if any unknown) in 1850 and another made for E. L. Percy, trunk dealer around the same time.[57] The eagles on the obverse of those tokens bear an image virtually identical to that found on Eagle (5A) identification discs marketed eleven years later.

It is also interesting to note that according to the Fuld brothers, who made an exhaustive study of Civil War tokens and store cards, all tokens minted by Scovill had a decorative border of beads around the obverse rim.[58] The Eagle (5A) identification disc is the only type observed with beads.[59] Because the Eagle (5A) is the only identification disc with beads rather than denticles or ropes, it is assumed they were made by Scovill. Scovill certainly had the equipment and expertise to mint and guild discs since throughout the war it manufactured huge numbers of buttons and equally prodigious numbers of patriotic tokens and store cards.

If further support is helpful, a relatively large number of Eagle (5A) discs were purchased by soldiers from New England, particularly by those from New Hampshire and Vermont. Perhaps Scovill was selling its discs just across the Connecticut border to soldiers in their mustering camps before they marched off to war.[60]

On the other hand, a number of patriotic

Left: A closeup of the die mark MERRIAM on the hanger tab of a Battles (4C) substyle. *Right:* A closeup of the die mark BOSTON on the hanger tab of a Battles (4C) substyle.

tokens bear a strong enough resemblance to the Eagle (5A) style identification disc so as to have possibly come from the same craftsman. Those with the greatest similarity to Eagle style discs were made by Shubael Davis Childs of 117½ Randolph Street, Chicago, Illinois. One is hard pressed to detect a significant difference between four of his patriotic tokens (Fuld nos. 278, 279, 280, and 280A) and the Displayed Eagle style identification discs.[61] Childs, like Scovill, also made sutler tokens and produced at least one for the 11th Wisconsin R.V. which had a displayed eagle similar to that found on the Eagle (5A) disc.[62]

There are only a few other identification disc styles that strongly resemble patriotic tokens.[63] The "Geo. B. McClellan" identification disc bears a striking resemblance to patriotic token done by Robert Lovett (McClellan 1A and Fuld 138A), while the "G.B. McClellan" style and the token cut by William H. Bridgens of 189 Williams Street, New York, New York, are very similar (McClellan 1B, Fuld 138).[64] Since we know that Frederick Smith left his name on those two McClellan discs, perhaps he gave the master dies away after McClellan's marketability had fallen, or maybe the token makers copied designs they found appealing. There is also a strong, but not overwhelming resemblance between the McClellan patriotic token done by F.C. Key and Sons of 329 Arch St, Philadelphia (McClellan 1C and Fuld 142), and to the small McClellan "Peninsular Campaign" identification disc (McClellan 1D and Fuld 142.)[65]

A merchant token manufactured by the Scovill Manufacturing Company.

The only other identification discs, the "Against Rebellion" and "War for the Union" shield styles (Shield 2A & 2B) for which a reasonable comparison can be made is with a series of patriotic tokens bearing a very similar shield and layout, but the words "Union Forever" and "1864" which were all cut by Charles D. Horter of 178 William Street, New York, New York. His own store card proclaimed that he was a die-sinker, and that he made medals. (Shield 2A & 2B and Fuld 339, 341, 342, 342A and 343).[66]

*****

The Union suffered more than 110,000 soldiers killed in action during the Civil War. Some of those who offered their "last full measure of devotion" were identified by the small discs which hung around their neck or glinted in the sun on a dusty and blood stained blue jacket. Undoubtedly many more would have been spared being counted among the missing in action had they purchased, or had the government supplied, some form of identification.

For those with the imagination to foresee the need, and the finances to be able to afford an identification disc, their comrades were able to either carry home or mail to grieving wives and families the medals that rose and fell with their loved one's final breath. Almost certainly, those discs became cherished keepsakes passed from wives to children, or parents to siblings.

Because of what it represented to its owner, and where it was worn, identification discs are one of the most personal and deeply significant Civil War relics a collector can acquire. When holding an identification disc, one can appreciate the courage, sacrifice and devotion of its wearer, and can almost feel the memories and pride it evoked in the wearer's golden years, or the love in which it was bathed when held by family members and descendants.[67]

Descriptions and Images of Identification Disc Styles

The identification disc styles and substyles included within the following catalog represent a select grouping of Civil War relics. To be included as an identification disc the medal had to meet a number of stringent criteria. First and foremost it had to be a disc, as distinguished from a shield shaped badge or geometric corps badge, regardless of whether biographical soldier information was included. The disc had to be metal and the images and words on the obverse (front) as well as the less common manufacturer-added wording or decorations on the reverse must have been placed on the disc by a minting process similar to that described in Chapter 1 rather than by casting, engraving or carving.[1]

Only Union discs are included because the authors know of no authentic minted Confederate identification discs like those worn by Union soldiers.

To be included in the catalog, some area of the disc had to have been set aside by the die-sinker as a place to insert information capable of supplying the identification of remains. Many soldiers wore patriotic tokens or silver coins around their necks either as an expression of patriotism or merely for safe keeping, sometimes with identification information scratched, engraved, or punched on one or both sides. Because those items were not intended primarily to provide identification, they are not included.

The manner in which the area for the information was set aside varies, but variations were not treated as exclusionary as long as apparently intended for the inclusion of information. On most discs, the reverse was left blank except for a raised rim and an embellishment of denticles, stylized ropes or beads just inside the raised rim. One type, however, had standard information, such as "Co." and "Reg." originally included when minted, with blank spaces for the rest of the information to be hand stamped by the sutler when purchased. Several other styles originally included lines above which the information was to be inserted. On a few, the area set aside for information consisted of a blank area incorporated into an image, such as an empty shield on the breast of an eagle. The key to designation as an identification disc, and inclusion, was that the manufacturer intended to preserve an area for identification type information.

Undoubtedly, there were coins and metal discs adapted for identification purposes during the Civil War, but examples of those types were not included in this work. They were as individual as the soldiers who had them made and are thus generally incapable of placement within a standardized category. Second, it is inherently difficult, if not impossible, for either an experienced or novice collector (including the authors) to distinguish between information stamped, engraved or scratched on an old coin or scrap of metal in 1862 from similar lettering which was added to the same type object by the same method in 1962 or 2007 and then aged.

One of the most difficult decisions was whether to include otherwise acceptable minted discs where the identifying information had been engraved, presumably by a jeweler, rather than stamped on by a sutler with a letter or number punch. In the end a compromise of sorts was reached. If a minted disc met all other criteria, it was not excluded solely because the information was engraved rather than stamped.[2]

Another issue that demanded extra deliberation was whether to include discs that met all other criteria but for which no example bearing stamped or engraved information has been observed. Opposing inclusion of such pieces was the position that if no example has been located with soldier's information, it is impossible to determine whether the item was intended to be used as an identification disc or merely as some sort of medal, such as for an election, or as adornment.[3] Arguing for inclusion was that such examples clearly look like they were intended as identification discs and they meet the other criteria for inclusion. It was determined to include them because of the fear that after publication a stamped or engraved version would surface from out of some farmer's field, a collector's safe, or from a flea market cigar box.

The authors also struggled over the designation of the type of metal from which the discs were minted. Most were minted from brass and then gilded with a thin layer of gold.[4] The few that were made from copper are also obvious. The problem arose mainly over those that were silver, white or gray in appearance. Those could have been made from any number of metals or alloys including pewter, base metal, silver, nickel, German silver, zinc, lead, or tin.[5] Unfortunately, neither of the authors have the expertise or the courage to subject their own silver or gray discs to the type of analysis necessary to determine the composition of the metal. Therefore, all white or gray discs have been designated as fabricated from "white metal," a generic attribution which includes all of the metals or alloys in the foregoing list.[6]

It is rare to encounter crossover discs, i.e., ones that have been listed as white metal here but which appear in brass or vice versa. Crossover pieces exist, usually made either as trial pieces or specially for the numismatic market that existed even in the early 1860s.[7] Crossover discs should not necessarily be avoided, but should be viewed with significant caution.

Finally, each substyle category listed includes an estimation of rarity compared to all other substyles reviewed. The ratings should not be taken as definitive, since they were based on a sample that consisted of only 615 stamped discs. There is no practical way to determine the number of all discs extant or the total number of any particular substyle.[8] Further, the reviewed discs were not accumulated for the purpose of gathering a scientific sample. One author selected his discs primarily on the basis of whether the soldier's unit fought at Second Bull Run, Antietam or Gettysburg. The other collected discs predominantly from Pennsylvania soldiers and a few for their scarcity. The largest collection observed was an amalgam of discs chosen on the basis of love, marketability, style variations, and those worn by soldiers from Connecticut.

The identification discs observed in online auctions probably represent the broadest and most random sample of the overall styles of discs extant, but how that presumption could be scientifically verified is beyond the authors' expertise.[9]

It would seem, though, that since it is rare to find more than a half dozen identification discs for sale at any single auction or show, or more than two or three for sale on-line in any given month, that the sample used should be fairly representative. Due to the limitations inherent in the size and method of the sample, no attempt will be made to provide an approximation of the total number of any one substyle as compared to the total extant identification discs.[10] Rather, the rating provided is strictly relative to the 615 discs appearing in the survey found in Chapter 5.

Using that list led to these classifications: *Common* equals over 100 similar substyles out of 615 samples available and listed in Chapter 5; *Scarce* equals 30 to 99 examples; *Rare* equals 10 to 29; *Very Rare* corresponds to five to nine examples; and *Extremely Rare* denotes four or fewer examples in the survey list.[11]

Identification Disc Catalog and Style Gallery

McClellan 1A
Obv.— Full bust left with "*Major General Geo B. McClellan * War of 1861" (brass/30mm/11g/die-sinker mark= Smith F, F. B. Smith, F. Smith B) *Rarity= Scarce*. [Dewitt/Sullivan #GMcC 1864-92.]
Rev.— Blank.[12]

McClellan 1B
Obv.— Full bust left with "*Major General G. B. McClellan *War of 1861" (brass, white metal/30mm/11g/ds= B Smith F). *Rarity= Rare.*[13]
Rev.— Blank.

McClellan 1C
Obv.— Small bust facing left in upper half with "Major General Geo. B. McClellan War of 1861"
(brass/30mm/10g). *Rarity= Scarce.* [Dewitt/Sullivan #GMcC 1864-93.]
Rev.— Blank.

2 • Descriptions and Images of Identification Disc Styles

McClellan 1D
Obv.— Small size, full bust left with "*Maj. Gen. G. B. McClellan*Peninsular Campaign"
(brass/19mm/3g). *Rarity= Rare.* [DeWitt/Sullivan #GMcC 1864-30.]
Rev.— Blank.

McClellan 1E
Obv.— Full Bust left with "Union and Liberty" with double lined rim (white metal/31mm/15g).
Rarity= Very Rare. [DeWitt/Sullivan #GMcC 1864-91.][14]
Rev.— Double lined border with space for information between.

2 • *Descriptions and Images of Identification Disc Styles*   31

McClellan 1F
Obv.— Full bust left with "Antietam Sep. 17th 1862" with double lined rim
(white metal/38mm/25g). *Rarity= Very Rare.*
Rev.— Double lined border with space for information or decorations between.

McClellan 1G
Obv.— Full bust left with "Antietam Sep. 17th 1862" with lined rim (white metal/38mm/26g). *Rarity= Extremely Rare.* [DeWitt/Sullivan #GMcC 1864-89.][15]
Rev.— Double lined border with space between for information and "CHEAT MOUNTAIN *** YORKTOWN WILLIAMSBURG FAIR OAKS GAINES MILL MALVERN HILL *** SOUTH MOUNTAIN" with the first and last curved around inside the interior border and the others centered.

2 • Descriptions and Images of Identification Disc Styles

McClellan 1H
Obv.— Full bust left with no legend (white metal/?/?). ***Rarity= Extremely Rare.***[16]
Rev.— Six stars. (Both courtesy Richard E. Burnham Collection)

McClellan II
Obv.— Bust ¼ right with "Maj. General McClellan" (brass or white metal/34mm/10g).
Rarity= Extremely Rare. [DeWitt/Sullivan #GMcC 1864-8 (A).][17]
Rev.— Double lined rim with "The war of 1861 & '62" & 31 Stars between rim and a row of beads.

McClellan 1J
Obv.— Bust ¼ right with "Maj. General McClellan" white metal/34mm/9g).
Rarity= Extremely Rare. [DeWitt/Sullivan #GMcC 1864-8 (B).]
Rev.— Rim with "The War of 1861, '62 & '63" and 13 Stars between the rope.

McClellan 1K
*Obv.—Bust ¼ right with "Maj. General McClellan" (white metal/34mm/9g).
Rarity= Extremely Rare. [DeWitt/Sullivan #GMcC 1864-8 (C).]
Rev.—"The War of 1861, 1862, 1863, & 1864" with wreath and Shield.

McClellan 1L
*Obv.— Full bust left with "Maj. Gen. Geo. B. McClellan U.S.A." (copper or bronze/31mm/13g).
Rarity= Extremely Rare. [DeWitt/Sullivan #GMcC 1864-12.][18]
Rev.— "First In The Hearts of His Soldiers" between rim and inner wreath.

McClellan 1Ma
a. *Obv.— Full bust left with "Geo. B. McClellan Major Gen. U.S.A." with double rim
(White Metal with silver wash/44.5mm/42g/ds= "WB" [possibly William Barber of the U.S. Mint]).
Rarity= Extremely Rare. [DeWitt/Sullivan #GMcC 1864-88.]
Rev.— Raised rim with additional rim line and "AMORE, PATRIAE, BELLIGERUM"
around upper half of rim.

2 • *Descriptions and Images of Identification Disc Styles* 39

McClellan 1Mb
*Obv.— Same as McClellan 1Ma. *Rarity = Extremely Rare.*
Rev.— Circular wreath of leaves and berries around open center.
[Not included in DeWitt/Sullivan.]

Shield 2A
Obv.— Full shield with "Union" in center and "Against Rebellion *1861*" around edge (brass/30mm/11g). *Rarity= Scarce.* [Dewitt/Sullivan #C 1861-11.]
Rev.— Blank.

Shield 2B
Obv.— Full shield with "War for the Union 1862"
(brass/28mm/9g/possibly with beads rather than denticles). *Rarity= Rare.*
Rev.— Blank.

Shield 2C
Obv.— One half sized shield in upper half with "The Union and the Constitution *War of 1861*"
(brass, white metal/30mm/10g). *Rarity= Rare.*
Rev.— Blank.

Shield 2D
Obv.— Full shield with crossed cannons and bayonets with "War of 1861 Union Forever" (white metal/31mm/12g). *Rarity= Extremely Rare.*
Rev.— Blank.

Washington 3A
Obv.— Full bust right with rope border inside rim and "George Washington. Born February 22, 1732." (white metal/31mm/14g/ds= Merriam or obliterated). *Rarity= Scarce*. [Baker #122T & U.][19]
Rev.— Blank with raised rim and rope or wreath instead of denticles and "War of 1861" and/or small decoration "-< >-" usually stamped on.

2 • *Descriptions and Images of Identification Disc Styles*   45

Washington 3B
Obv.— Full bust right with "Union [below]" & 34 stars around border (brass/32mm/14g).
Rarity= Rare. [Baker #620A.]
Rev.— Blank.

Washington 3C
Obv.— Full bust right wit "Union [below]" and border of 34 stars (brass/32mm/14g).
Rarity= Scarce.[20] [Baker #620.]
Rev.— "Co. _, ___Reg.___, Volunteers, Enterd [sic] Service, ___, 186_" with space around periphery for name and blanks for other soldier information.

2 • *Descriptions and Images of Identification Disc Styles* 47

Washington 3D
Obv.— Full bust right with "George Washington [above]" & "Security [below]"
(white metal/32mm/12g/ds= "Lovett" and "R.L."). ***Rarity= Very Rare***.
[Baker #621A, obverse only. See Chapter 4.]
Rev.— Blank.[21]

Washington 3E
*Obv.— Full bust right with border of leaves, and "George Washington First President of the Ud. States" (white metal/35mm/?/ds= G.H. Lovett N.Y.). **Rarity= Extremely Rare.** [Baker #113G.]
Rev.— Blank. (Both courtesy Richard E. Burnham Collection)

2 • *Descriptions and Images of Identification Disc Styles* 49

Battles 4A
Obv.—"Fought in Battles*" [above] and "1861, 2 & 3 United States [below]" with eight lines for battles (brass/28mm/10g/rope instead of denticles). *Scarcity= **Extremely Rare**.*
Rev.— Blank.

Battles 4B
Obv.— Rope border between rim and wreath with "War of 1861 [above]" and "Engaged in Above Battles [below]" (white metal/31mm/12g/), usually seen without a suspension hole). *Rarity= Rare.*
Rev.— Blank. (Both courtesy John Craig Collection)

Battles 4C
Obv.— Rope inside rim with "War of 1861 [above]" and "Engaged in the Above Battles [below]"
and a hanger tab. (white metal/31mm round/12g//ds= "Merriam"
on one side of the base of the hanger tab and "Boston" other]). *Rarity= Rare.*
Rev.— Blank.

Battles 4D
*Obv.—"*FOUGHT IN BATTLES*" [above] and "1861, 2 & 3 UNITED STATES" below with eight blank lines in between (brass/27mm/?). *Rarity= Extremely Rare.*
Rev.— Twelve blank lines for soldier's identification and/or additional battles.
(Both courtesy Richard E. Burnham Collection)

2 • Descriptions and Images of Identification Disc Styles

Eagle 5A
Obv.— Displayed Eagle with shield on chest and "War of 1861 [above]" and "United States [below]" (brass/28mm/9g/beads instead of denticles).[22] *Rarity= Common* [169 out of 615].
Rev.— Blank

Eagle 5B
Obv.— Displayed Eagle with shield and "Union [between wings] In The War of 1861, 2 & 3" (Brass with silver wash/30mm/10g). ***Rarity= Very Rare.***
Rev.— Blank.

Eagle 5C
Obv.—Displayed Eagle with shield and "Union [between wings] In The War of 1861, 2 & 3"
(Brass/30mm/10g). ***Rarity= Rare.***
Rev.—"In Battle Of."

Eagle 5D
Obv.— Displayed Eagle with shield and "Union [between wings] In the War of 1861, 2 & 3, & 4" (brass w/silver wash/30mm/10g). *Rarity= Extremely Rare.*
Rev.— Blank, possibly with beads.

Eagle 5E
*Obv.— Displayed Eagle with shield but higher wing tips and beak than above and "THE UNION MUST AND SHALL BE PRESERVED" (brass/24mm/?). *Rarity= Extremely Rare.*
Rev.— Blank with neither raised border nor denticles.

Lincoln 6A
Obv.— Full bust right with "Abraham Lincoln President U.S.*War of 1861*"
(brass or white metal/30mm/10g/ds= F B Smith). ***Rarity= Rare.***[23]
Rev.— Blank.

Banks 7A
Obv.— Full bust right with "N, P, Banks Maj, Gen, U,S,A," (white metal/34mm/17g).
Rarity= Extremely Rare.
Rev.— Blank.

Union 8A
Obv.—"No Compromise with Armed Rebels [around edge]" "UNION [center]" surrounded by "May the Flourish" (white metal/32mm/8g). *Rarity= Extremely Rare.*[24]
Rev.—Blank.

Grant 9A
Obv.— Full bust right circled by "Lieu. Gen. U.S. Grant and 13 stars" with wreath around edge. (brass or white metal/40mm/36g or 13g). *Rarity= Extremely Rare.* [DeWitt/Sullivan #USG 1868-5.]
Rev.— Shield topped by Eagle and surrounded by 10 partially furled flags.

Grant 9B
Obv.— Full bust left (primitive) circled by "UNCONDITIONAL SURRENDER"
with hanger tab (copper/22mm/?). ***Rarity= Extremely Rare.***[25]
Rev.— Blank. (Both courtesy Richard E. Burnham Collection)

Grant 9C
*Obv.— Full bust left in circle with "LIEUT. GEN. ULYSSES S. GRANT U.S.A." and "*1864*" below bust. (copper/31mm/?). *Rarity= Extremely Rare.* [DeWitt/Sullivan #USG 1864-1, *obverse only.*][26]
Rev.— Blank. (Both courtesy Richard E. Burnham Collection)

Sigel 10A
Obv.— Bust ¼ left with "Maj. Gen. Franz Sigel Missouri * Virginia" (brass/19mm/3g).
Rarity= Very Rare.
Rev.— Blank.

Hooker 11A
Obv.— Bust ¾ left with "Maj. General Joseph Hooker" and hanger tab
(white metal or brass/22mm/4g). Rarity= *Extremely Rare*.
Rev.— "Hooker's Old Division Old Guard."

Scott 12A
*Obv.— Full bust left with "MAJ. GENERAL WINFIELD SCOTT" (brass/26mm/?).
Rarity= Extremely Rare. [DeWitt/Sullivan #WS 1852-13, obverse only.]
Rev.— Blank. (Both courtesy Richard E. Burnham Collection)

2 • Descriptions and Images of Identification Disc Styles

Engineer 13A
Obv.— Engineers Castle (brass/19mm/?). *Rarity= Extremely Rare.*
Rev.— Blank. (Both courtesy Richard E. Burnham Collection)

New York 14A
*Obv.— New York State Seal with "UNION & CONSTITUTION" above and banner below (white metal/40mm/?). **Rarity= Extremely Rare.**
Rev.— "N.Y.S. Vols" [period under last s] and shield circled by garlands & and banner with "1776-1861." (Both courtesy Richard E. Burnham Collection)

2 • *Descriptions and Images of Identification Disc Styles*   69

Lady Liberty 15A
*Obv.—Seated Lady Liberty Extending Wreath with "Honor Is the Reward of Loyalty" (white metal/31mm/12g/ds= "RL" [Robert Lovett Jr.]). **Rarity= Extremely Rare.**
Rev.—"War of 1861" at top with a wreath around the edge.[27]

Double Blank 16A
Obv. & Rev.— Blank with raised rim and inside ridge and hanger tab (white metal/25mm/6g).
Rarity= Very Rare.

2 • *Descriptions and Images of Identification Disc Styles*   71

Sherman 17A
*Obv.—Bust ¾ left surrounded by circle with scrolls above and "MAJ. GEN. Wm. T. SHERMAN. U.S.A" (brass/30mm/?). Rarity= **Extremely Rare**.*
Rev.— Blank.

Sherman 17B
Obv.— Bust ¾ right surrounded by wreath with "MAJ. GEN. W. T. SHERMAN."
(white metal or brass/31mm/9g). *Rarity= Extremely Rare.*
Rev.— Six pointed compass rose with circular area in center for soldier information.

2 • Descriptions and Images of Identification Disc Styles

Union League 18A
Obv.— Shield in center with banner draped over bearing "E Pluribus Unum" and concentrically from top edge stars, "UNION LEAGUE" "July 4th" "1776 1863" and "Philadelphia" (White metal/34mm/17g). ***Rarity= Extremely Rare.*** [DeWitt/Sullivan #U 1862-2]
Rev.— Blank.

Rhode Island 19A
Obv.— Stylized Rhode Island State Seal in center with a shield around "HOPE" above anchor and "1844" below and 13 stars above shield. (White metal, copper/29mm/?). ***Rarity= Extremely Rare.***
Rev.— Wreath around edge with "WAR OF 1861" above between ends of wreath and "ENGAGED IN THE ABOVE BATTLES" inside of wreath.

All Seeing Eye 20A

*Obv.— Eye in center surrounded by rays with "THE UNION MUST AND SHALL BE PRESERVED" around edge. (White metal, brass, copper/22mm/?). **Rarity= Extremely Rare.** [Dewitt/Sullivan #C 1861-8.][28] [No image available, obverse similar to the Fuld #228 patriotic token by J. Lovett.]
Rev.— Blank

③

SOLDIERS AND THEIR IDENTIFICATION DISCS

In this chapter are presented more than 100 identification discs that belonged to soldiers in the Union Army. These soldiers represent the many thousands that fought for the Union. A short biography of each soldier is included with the pictures of his disc. These biographies run the gamut from death in combat to permanent wounds to deserters. But all of the men served their country in some form for some period.

Connecticut

Private John Scott

COMPANY B, 8TH CONNECTICUT VOLUNTEERS

Mustered in September 21, 1861, at Hartford, Connecticut, for 3 years, age 18, as a private; born in Thompsonville, Connecticut.

5' 7" tall, light blue eyes, dark brown hair, dark complexion, was a mechanic.

Not stated November 1861 to February 1862.

In hospital March 22, 1862.

Listed as present from March 1862 to March 1863.

Appointed corporal on March 24, 1863.

Present from April 1863 to December 1863. Re-enlistment as Veteran Volunteer on December 23, on 30 day furlough January 1864.

Left: Obv. Pvt. Scott. *Right:* Rev. Scott.

Present from February 1864 to August 1864.
Appointed Sergeant in August 1864.
Wounded on September 29, 1864, at Chaffins Farm; absent from September 1864 to October 1864 in hospital (sick/wound).
Present from November 1864 to October 1865.
Mustered out on December 12, 1865.
Saw action at Roanoke Island, Fort Macon, South Mountain, Antietam, Fredericksburg, Suffock, Fort Huger, Dismal Swamp, Port Walthall, Swift Creek, Fort Darling Cold Harbor, Petersburg, Chaffin Farm.

Delaware

Private Samuel Mays

COMPANY E, 2ND DELAWARE INFANTRY

Enlisted on June 15, 1861, at age 18 at Wilmington, Delaware, for 3 years.
5' 6½" tall, blue eyes, sandy hair.
Stage driver from Blackhorse, Pennsylvania.
Present March 1862 until June 1862.
Absent in hospital July/August 1862.
Discharged for general disability on October 22, 1862.
Died May 27, 1904.
Saw action at Gaines's Mill, Savage's Station, Peach Orchard, White Oak Swamp and Malvarn Hill.

Left: Obv. Mays. *Right:* Rev. Mays.

Illinois

Private James H. Phillips

COMPANY A, 12TH ILLINOIS CAVALRY

Mustered at Springfield, Illinois, on February 28, 1862, age 17, for 3 years as a private.
Born in New York, worked as a farmer.
5' 3" tall, grey eyes, light hair.

Left: Obv. Phillips. *Right:* Rev. Phillips.

Listed as present from March/April 1862 to March 1863.
Absent, wounded on picket duty, April 11, 1863.
In hospital, Belle Plain, Virginia, slight wound of right foot.
Present May/June 1863.
Absent August 1863 to September/October 1863.
Shown at dismounted camp, Washington D.C., on August 5, 1863.
Reenlisted on February 28, 1864, as Veteran Volunteer at St. Louis.
March/April 1864 sick since April 19 at Port Hudson.
Detached service as orderly May/June 1864.
Present July/August 1864 to July/August 1865.
Absent on furlough September/October 1865 in Chicago.
Report absent without leave December 16, 1865.
Mustered out March 17, 1866, at New Orleans.
Saw action at Norfolk, Fair Oaks, Oak Grove, Seven Days, 2nd Bull Run, Chantilly, Fredericksburg, Chancellorsville, Gettysburg.

Indiana

Private George Bickle

Company C, 9th Indiana Infantry

Enlisted Elkhart, Indiana, April 19, 1861, for 3 months, mustered out July 29, 1861.
Re-enlisted at Wakarusa, Indiana, February 26, 1862, for 3 years, at age 27; 6' 4" tall; brown eyes and hair. Born in Preble County, Ohio, and occupation was farmer.
Present for duty from March 1862 to April 1862.
Absent at hospital May 1862 to medical discharge September 30, 1862.
Saw action at Philippi, West Virginia, Laurel Hill, Bealington, Corrick's Ford.

Private Henry O'Blenis

Company C, 20th Indiana Infantry

Mustered in at Lafayette, Indiana, on July 22, 1861.
For 3 years, age 20 or 25?, as a private.
5' 7" tall, hazel eyes, red hair, light complexion.

Left: Obv. Bickle. *Right:* Rev. Bickle.

Not stated July/August to November/December 1861.
Listed as present from January 1862 to March/April 1862.
Absent July/August 1862.
Wounded in action June 25 or 26, 1862. Wound of left leg above kneecap.
At Mount Pleasant USA to general hospital, Washington, D.C., May to August 1862.
At USA G.H. Portsmouth Grove, Rhode Island, September/October 1862.
Absent September/October 1862 to July 23, 1863.
Assigned to extra duty as nurse to February 28, 1863.
July 23, 1863, transferred to 39th Company 1 Battalion Invalid Corps.
Present July/August, 1863 detached September/October, 1863 to Kingston, New York.
Mustered out July 28, 1864.
Died March 22, 1904.
Saw action at Norfolk, Fair Oaks, Oak Grove, Seven Days.

Left: Obv. O'Blenis. *Right:* Rev. O'Blenis. The meaning and dating of the stamped "No 00047" is unknown.

Sergeant John Dickey

Company H, 20th Indiana Infantry

Enlisted at Lafayette, Indiana, on July 22, 1861, for 3 years, age 18.
Born in Scotland, worked as a farmer.

5' 9" tall, grey eyes, dark hair, light complexion.
Listed as present from January 1862 to February 1864.
Promoted to corporal July/August 1862.
Promoted to sergeant May/June 1863.
Promoted to 1st sergeant on December 15, 1864.
Lost on march between April 21 and May 6, 1863, one knapsack, value $2.14.
Discharged near Culpepper, Virginia, on February 20, 1864.
Enlisted as Veteran Volunteer on February 21, 1864, for 3 years, as a sergeant, later 1st sergeant.
Present until March 1865.
Absent sick in Philadelphia March and April 1865.
Present May to July 1865.
Mustered out July 12, 1865, at Jeffersonville, Indiana.
Saw action at Norfolk, Fair Oaks, Oak Grove, Seven Days, 2nd Bull Run, Chantilly Fredericksburg, Chancellorsville, Gettysburg, Locust Grove, Mine Run Wilderness, Todd's Tavern, Po River, Spotsylvania, Totopotomy, Cold Harbor, Deep Bottom, Strawberry Plains, Petersburg.

Left: Obv. Dickey. Sergeant Dickey's identification disc was dug up near Fredericksburg. *Right:* Rev. Dickey.

Maine

Corporal Samuel R. Bishop

Company B, 17th Maine Infantry

Mustered in at Portland, Maine, on July 25, 1862, for 3 years, age 19, as a private, worked as a clerk.
5' 5" tall, black eyes, black hair dark complexion.
No record for July/August 1862.
Listed as present from September 3, 1862, to April 1863.
Sent to hospital June 10, 1863.
In Mansion House Alexandria Hospital to February 1864.
Present March 1864 to June 1865.
March 1864 stoppage of $11.23 for transportation.
Promoted to corporal July 1, 1864.
July/August 1864 stoppage of 62 cents for ordnance.
January/February stoppage $23.76 for transportation.

Promoted to sergeant January 31, 1865.
Mustered out June 4, 1865, near Washington, D.C.
Died January 8, 1901.
Saw action at Fredericksburg, Chancellorsville, Wilderness, Spotsylvania Court House, Bloody Angle, Cold Harbor, Petersburg, Appomattox Court House.

Left: Obv. Bishop. *Right:* Rev. Bishop.

Maryland

Private Thomas Trac(e)y

Company F, 2nd Maryland Infantry

Records show name as both Tracy and Tracey.
Mustered in on August 8, 1861, at Herrford, Maryland, for 3 years, age 22.
5' 9" tall, with hazel eyes, black hair, dark complexion, was a laborer.
Not stated August to December 1861.
Listed as present from January 1862 to April 1863.
Noted on April 23, absent with out leave 17 days.
Present May 1863 to December 1863.
Mustered out on December 31, 1863, to allow re-enlistment in Veteran Volunteer.

Left: Obv. Trac(e)y. *Right:* Rev. Tracy. Private Tracy had the battle "Antietam" stamped on his identification disc, which is unusual on McClellan 1A substyle.

Present January and February 1864.
Absent March and April on furlough.
Present May and June 1864.
Absent July/August 1864.
Died September 26, 1864, at home.
Saw action at 2nd Bull Run, Chantilly, South Mountain, Antietam, Fredericksburg, Knoxville campaign, Spotsylvania, North Anna River, Cold Harbor.

Massachusetts

Sergeant John W. Cullinan

Company B, 9th Massachusetts Infantry

Enlisted on June 11, 1861, at Boston, Massachusetts, for 3 years, age 29.
5' 7" tall, hazel eyes, light hair, was a shoemaker.
Present from January 1862 to June 1862.
Wounded and captured June 27, 1862, at Gaines' Mill.
Prisoner until July 25, 1862, in hospital until November 5, 1862.
Present to mustering out June 21, 1864, and promoted to sergeant September 1863.
Enlisted Veteran Reserve Corps, August 27, 1864.
Discharged November 17, 1865.
Died April 30, 1922.
Saw action at Hanover Court House, Gaines' Mill, Chancellorsville, Gettysburg, Rappahannock Station, Mine Run, Wilderness, Laurel Hill, North Anna River, Bethesda Church, and Cold Harbor.

Left: Obv. Cullinan. *Right:* Rev. Cullinan. Sergeant Cullinan may have had his rank added after his promotion based on the unusual placement of "SERG.T."

Private Herman Donath

Company K, 19th Massachusetts Infantry

Mustered in August 28, 1861, at Lynnfield, Massachusetts, age 19, for 3 years.
Present from January 1862 to January/February 1863.
Promoted to sergeant September 1, 1862.

Promoted to sergeant major October 14, 1862.
Promoted to 2nd lieutenant December 14, 1862, transferred to Company I.
March/April 1863 absent on leave, returned April 10.
Promoted to 1st lieutenant May 1, 1862, transferred to Company C.
Killed in action July 3, 1863.
Saw action at Hanover Court House, Gaines' Mill, Chancellorsville, Gettysburg.

Right: Lieutenant Donath, from the Massachusetts MOLUS Collection (courtesy U.S. Army Military History Institute).

Left: Obv. Donath. *Right:* Rev. Donath. The reverse of Sgt. Major Donath's identification disc probably has a die error in that it is flat with no rim marks.

Michigan

Private Daniel Ferguson

COMPANY I, 2ND MICHIGAN INFANTRY

Mustered in on May 25, 1861, at age 27, for 3 years at Detroit.
Not stated May to November/December 1861.
Present from January 1862 to July/August 1862.
Not stated September/October 1862.
Present November/December 1862 to January 30 1864.
Assigned to 17th Michigan January 30, 1864.
Returned to 2nd Michigan April 20, 1864.
Absent May/June with leave in Detroit.
Mustered out July 21, 1864.
Died July 1, 1920.
Saw action at First Bull Run, Yorktown, Williamsburg, Glendale, Seven Pines, Malvern Hill, Second Bull Run, Chantilly, Ferdericksburg, Vicksburg, Jackson, Blue Springs, Campbell's Station, Knoxville, Wilderness, Spotsylvania, North Anna River, Cold Harbor, Petersburg, Weldon Railroad, Poplar Spring Church, Ream's Station, Hatcher's Run.

Left: Obv. Ferguson. *Right:* Rev. Ferguson. Private Ferguson left his home town of "BOWMANSVILLE, NY.," which he preferred on his identification disc even though he served in a Michigan regiment.

Private Alfred Ames

Company I, 3rd Michigan Infantry

Mustered in on June 10, 1861, at Grand Rapids, Michigan, age 21.

5' 9" tall, blue eyes, light hair, and light complexion, was a farmer.

Not stated June 10 to November/December 1861.

Present January 1862 to May/June 1862.

July/August to November/December listed as "Daily Duty General Hostler."

Present January/February 1863 to July/August 1863.

Note "stoppage of $10 by order of Lieutenant Colonel Pierce for straggling on the march" July/August 1863.

Absent, sick in hospital, October 12, 1863.

At 3rd division USA general hospital (Washington Street Branch), Alexandria, Virginia.

Present in hospital October 13 to March 18, 1864.

Transferred to Augur U.S.A. General Hospital near Alexandria, Virginia.

Discharged June 10, 1864, at end of three year enlistment.

Left: Obv. Ames. *Right:* Rev. Ames. Private Ames scratched "Bull Run" on his identification disc twice in sequence to show he was at both these battles, apparently, because the sutler neglected to include them in the list of stamped battles.

Last paid February 29, 1864; due the government $8.86 for clothing allowance, soldier due $100 bounty; left Washington on June 8 for Detroit.

Died August 31, 1929.

Saw action at Blackburn's Ford, First Bull Run, Yorktown, Williamsburg, Fair Oaks, Savage's Station, Peach Orchard, Charles City Cross Roads, Malvern Hill, Second Bull Run, Chantilly, Ferdericksburg, Chancellorsville, Gettysburg.

New Hampshire

Private George L. Chase

COMPANY H, 2ND NEW HAMPSHIRE VOLUNTEER INFANTRY

Mustered in on June 5, 1861, at Portsmouth, New Hampshire, for 3 years, age 18.

Not stated June 1861–April 1862.

Detailed as teamster August 31, 1861.

Listed as present from April 1862 to May 1863.

Absent sick May 25, 1863, in Concord, New Hampshire.

Apprehended as deserter August 14, 1863, in New Hampshire, fined $3.50.

Present November 1863–June 1864.

Mustered out on June 21, 1864, at Concord New Hampshire.

Mustered in Company C 18th New Hampshire Volunteers on September 13, 1864, as private.

Present October 1864 — February 1865.

Absent in hospital March 1865, admitted to Slough General Hospital, Alexandria, April 26, 1865, chronic diarrhea.

Transferred to Manchester, New Hampshire, May 22, 1865.

Mustered out on June 10, 1865.

Died June 15, 1865.

Saw action at 1st Bull Run, Yorktown, Williamsburg, Fair Oaks, Malvern Hill, 2nd Bull Run, Fredericksburg.

Left: Obv. Chase. *Right:* Rev. Chase. Private Chase had the sutler stamp "WAGONER" on his identification disc to show his assignment.

Private Paul Morgan

Company E, 2nd New Hampshire Volunteer Infantry

Mustered in on September 17, 1861, at Keene, New Hampshire, for 3 years, age 22; born in Clinton, Massachusetts, was a farmer.

5' 7" tall, blue eyes, light hair, light complexion.

Not stated September 1861/February 1862.

Present March 1862 to August 31, 1862.

Wounded August 29, 1862, absent September 1862 to November 10, 1862.

In hospital, Annapolis, Maryland.

Loss of right arm, amputated middle of humeric, because of gunshot wound.

Enlisted in 99th Company 2nd Battalion Invalid Corps on July 14, 1863.

Discharged January 22, 1864.

Died March 5, 1922.

Saw action at Yorktown, Williamsburg, Fair Oaks, Malvern Hill, 2nd Bull Run.

After the loss of his arm, a very faded image of Private Morgan was included in his pension file, now in the National Archives.

Left: Obv. Morgan. *Right:* Rev. Morgan.

Private Joseph H. Foss

Company A, 5th New Hampshire Volunteer Infantry

Mustered in on August 9, 1864, at West Lebanon, New Hampshire, as a substitute, for 3 years, age 21, was a laborer.

5' 4½" tall, grey eyes, brown hair, dark complexion.

Substituted for Natham H. Patch.

Present September/October 1864.

Discharged November 1, 1864, for disability, hernia; had disability before enlistment.

Not entitled to transfer to Veteran Volunteers.

Saw action at Poplar Springs Church and Hatcher's Run?

Left: Obv. Foss. *Right:* Rev. Foss.

Private James Kirby
Company I, 6th New Hampshire Volunteer Infantry

Enlisted on December 3, 1861, for 3 years, age 18.
Not stated December 1861.
Listed as present from January 1862 to April 8, 1863.
Reported wounded on December 13, 1862, at Fredericksburg.
Absent in hospital in Lexington, Kentucky, April 8, 1863 until December 1863.
Mustered out on January 19, 1864, to allow re-enlistment in Veteran Volunteers.
Present January 1864 to June 1864.
Reported wounded May 12, 1864, at Spotsylvania Court House.
Absent in hospital June 7, 1864.
Present November 1864 to muster out on July 17, 1865.
Alive in 1891 based on reference in Regimental History.
Saw action at 2nd Bull Run, Chantilly, South Mountain.
Antietam, Fredericksburg, Vicksburg, Wilderness, Spotsylvania CH, North Anna River, Cold Harbor, Petersburg, Weldon Railroad, Hatcher's Run, Appomattox Court House.

Left: Obv. Kirby. *Right:* Rev. Kirby. The sutler misspelled Private Kirby's name as "JERBY."

Private Charles L. Perry
Company D, 9th New Hampshire Volunteer Infantry

Mustered in August 9, 1862, at Conway, New Hampshire, for 3 years, age 18, as a private.
Listed as present from August 23 1862, to February 1863.
Absent March/April 1863 listed as "Deserted on the march from Boonsborough, Kentucky April 17, 1863, since arrested and confined Richmond, Ky."
Absent May/June 1863, held by provost guard awaiting sentence of court-martial.
Present July/August and September/October under arrest.
Absent November/December, sick in post hospital.
Absent January/February 1864, in hospital at Paris, Kentucky.
Absent March/April and May/June, in hospital at Annapolis, Maryland.
Present with company July/August 1864.
Absent September/October 1864, listed as "missing in action."
Listed as POW September 30, 1864, at Poplar Grove Church, Virginia.
On POW form is note "joined rebel army while a POW at Salisbury, N.C. no date."
Muster out roll dated June 10, 1865, lists him as due $75 bounty.
Saw action at South Mountain, Antietam, Fredericksburg, Petersburg, Poplar Grove Church.

Left: Obv. Perry. *Right:* Rev. Perry.

Private Ara W. Adams
Company F, 9th New Hampshire Volunteer Infantry

Mustered in August 5, 1862, at Concord, New Hampshire, for 3 years, age 18(?), as a private.
Listed as present from September/October 1862 and November/December 1862.
Absent January 24, 1863, in hospital.
April 28, 1863, sick in Concord New Hampshire.
July/August 1863 noted sick in hospital in Covington, Kentucky.
Due government $9.22 for transportation.
Muster out roll dated June 10, 1865, lists him as due $75 bounty.
Last paid on October 31, 1862.
Father in same regiment.
Died April 25, 1897.
Saw action at South Mountain, Antietam, Fredericksburg.

Left: Obv. Adams. *Right:* Rev. Adams. Private Adams' father, Sylvanus, is reported to have served in the same regiment as the "principal musician."

Private Stephen O. Gray

COMPANY C, 12TH NEW HAMPSHIRE VOLUNTEER INFANTRY

Mustered in on September 5, 1862, at Alexandria, New Hampshire, age 27 for 3 years.
5' 7½" tall, black eyes, brown hair, fair complexion, was a farmer at enlistment.
Bounty of $25, advance of $13.
Present from September 1862 to July/August 1863.
Absent in hospital with slight wound of right calf and diarrhea since July 2, 1863, at Gettysburg.
At Point Lookout, Maryland, hospital July 22, 1863.
Issued 1 pair pants, 1 pair drawers, 1 woolen shirt, 1 blouse, and 1 pair of soxes [sic] at hospital.
Furloughed August 7 to 26, 1863.
Discharged for disability October 16, 1863.
Accountable for 1 Springfield rifle and appendages,
1 set accoutrements, 1 knapsack, 1 haversack and 1 canteen.
Died of chronic diarrhea at New Durham, New Hampshire, on February 13, 1864.
Saw action at Fredericksburg, Chancellorsville, Gettysburg.

Left: Obv. Grey. *Right:* Rev. Grey.

Private Theodore Sanborn
Company D, 12th New Hampshire Volunteer Infantry

Enlisted on September 5, 1862, at Sanbornton, New Hampshire, age 21.
5' 6½" tall, blue eyes, dark hair, a farmer at enlistment.
Present from September 1862 to July 1864.
Absent in hospital August 18, 1864, to September 1864, returned to duty, but returned to hospital November 20, 1864.
Died of chronic diarrhea at Fort Monroe Hospital January 28, 1865.
Saw action at Fredericksburg, Chancellorsville, Gettysburg, Swift Creek, Relay House, Drewry's Bluff, Port Walthall Junction, Cold Harbor, and Petersburg.

Left: Obv. Sanborn. *Right:* Rev. Sanborn.

Private Charles L. Sweatt
Company F, 12th New Hampshire Volunteer Infantry

Enlisted on August 14, 1862, at Pittsfield, New Hampshire, age 26?
5' 4½" tall, dark eyes, dark hair, light complexion.
Present from September 1862 to April 20, 1863.
Absent in Division Hospital April 1863 to November 1863 returned to duty.

Left: Obv. Sweatt. *Right:* Rev. Sweatt.

Present November 1863 to May 1864, detached service at Brigade Headquarters.
Returned to hospital June 25, 1864.
At hospital until discharge June 21, 1865.
Died November 11, 1911.
Saw action at Fredericksburg, Swift Creek, Relay House?

Private Joshua Janvrin
Company D, 14th New Hampshire Volunteer Infantry

Enlisted on September 23, 1862, at Seabrook, New Hampshire, for 3 years, at age 21.
Present from September/October 1862 to May/July 1864.
Not stated July/August 1864.
Absent September/October, sick at U.S. General Hospital, Sandy Hook, Maryland.
Transferred to 10th Reg V.R. Corps October 18, 1864.
Shown on muster out roll dated July 8, 1865
Discharged on June 28, 1865, according to New Hampshire adjutant report.
Saw no action.

Private Sweatt, from the *History of the Twelfth Regiment New Hampshire Volunteers in the War of the Rebellion* (Concord, New Hampshire: Ira C. Evans, Public Printer, 1897).

Left: Obv. Janvrin. *Right:* Rev. Janvrin.

Private J. Brown
Company G, 15th New Hampshire Volunteer Infantry

Mustered in on October 11, 1862, for nine months.
Mustered out with regiment on August 13, 1863.
Saw action at Port Hudson, Louisiana.

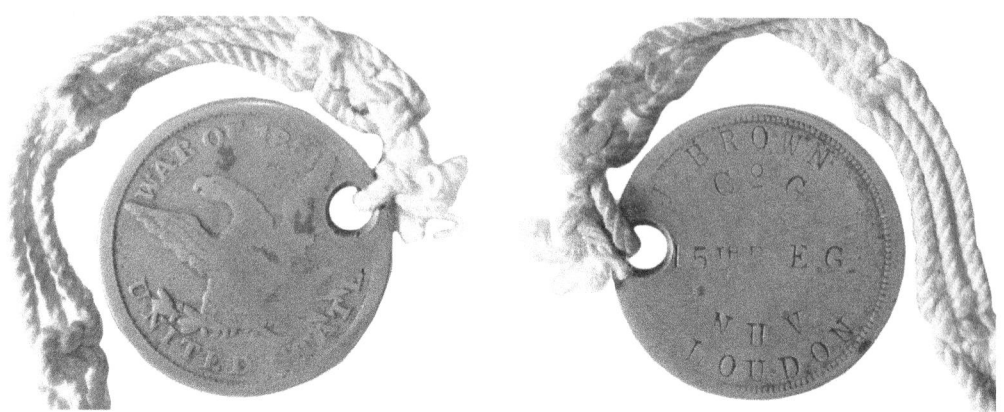

Left: Obv. Brown. *Right:* Rev. Brown.

New Jersey

Private David Dilts

COMPANIES A AND E, 5TH NEW JERSEY VOLUNTEER INFANTRY

Mustered in on August 14, 1861, in Company A.
Transferred to Company E on August 11, 1862.
Mustered out on September 7, 1864.
Saw action in the Peninsular Campaign, Bristoe Station, 2nd Manassas, Chantilly, Chancellorsville, Gettysburg, Overland Campaign.

Left: Obv. Dilts. *Right:* Rev. Dilts.

Private Elmer Errickson

COMPANY B, 10TH NEW JERSEY VOLUNTEER INFANTRY

Mustered in October 22, 1861, as a Private in Company B at Hiserville, New Jersey, for 3 years, age 20, as a private.
Born in Cumberland County, New Jersey.
5' 7½" tall, gray eyes, light hair, light complexion, was a farmer.
Not stated January 1862 to December 1862.

Shown as rejoined company June 3, 1862.
Listed as present from January 1863 to June 1864.
Listed as corporal as of July/August (June 8?) 1863.
Reenlisted on February 24, 1864.
Wounded on June 1, 1864, at Cold Harbor.
In First Division General Hospital Alexandria, Virginia.
Compound fracture of anterior part of right parietal bone plus gunshot flesh wound of left hand, ball extracted.
Died in hospital on June 8, 1864.
Buried on June 9, 1864, in grave # 2071 in military cemetery, Alexandria, Virginia.
Effects shipped to father, received on June 25, 1864.
Saw action at Wilderness, Cold Harbor.

Corporal Errickson's grave number in the Alexandria National Cemetery is 2071.

Left: Obv. Errickson. *Right:* Rev. Errickson.

Private Neal McDevett (McDevitt)

COMPANY D, 11TH NEW JERSEY VOLUNTEER INFANTRY

Mustered in on September 27, 1864, as a substitute.
He was age 20; 5' 7" tall; green eyes, blonde hair, a native of Ireland, and was employed as a boatman.

Left: Obv. McDevett. *Right:* Rev. McDevett.

He was present for duty from September 1864, until mustered out on June 6, 1865.
Fined $10.00 for straggling.
Saw action at Petersburg.

New York

Corporal George F. Beach
Company C, 22nd New York State Volunteer Infantry

Enlisted on June 6, 1861, at Troy, New York, for 2 years, age 23, as a corporal.
5' 9½" tall, blue eyes, light hair, salesman from Essex.
Not stated July 1861 to December 1861.
Present from January 1862 to August 29, 1862.
Reported missing since fight at 2nd Bull Run.
Present September 1862 to November 1862.
Absent sick until muster out June 19, 1863.
Enlisted September 7, 1863, as a private.
Company I 2nd Regiment New York Veteran Cavalry.
Promoted to 2nd lieutenant November 10, 1863.
Present January and February 1864.
Not stated March/April 1864.
Present May to August 1864.
September/October 1864 scout.
Present November 1864 to November 1865.
Promoted to 1st lieutenant January 15, 1865.
Mustered out November 8, 1865.
Died January 29, 1884.
Saw action at 2nd Bull Run, South Mountain, Antietam.

Left: Obv. Beach. *Right:* Rev. Beach.

Private William Livingston
Company D, 22nd New York State Volunteer Infantry

Mustered on June 6, 1861, at Troy, New York, for 2 years, age 23, as a private.
Not stated July 1861 to December 1861.

Present from January 1862 to July/August 1862.
September/October 1862 on extra duty, ambulance driver.
Absent September/October 1862.
On detached service ambulance corps.
Mustered out June 19, 1863, at Albany, New York.
Last paid February 28, 1863.
Due soldier $5.45 for clothing allowance.
Saw action at 2nd Bull Run, South Mountain?, Antietam?

Left: Obv. Livingston. *Right:* Rev. Livingston.

Assistant Surgeon Isaac Welsh

25TH NEW YORK STATE VOLUNTEER INFANTRY

Mustered in December 16, 1862, at Albany, New York, for 3 years, age 39, as 1st lieutenant.
Listed as present from December 1862 to April 26, 1863.
Resigned and honorably discharged.
Died June 23, 1878.

Left: Obv. Welsh. The listed battles took place before Assistant surgeon Welsh joined the 25th New York, although he may have been a contract surgeon before his commissioning. *Right:* Rev. Welsh. Assistant surgeon Welsh had his rank, "ASS'T SURG," stamped on his identification disc.

Sergeant George W. Steele

COMPANY G, 30TH NEW YORK STATE VOLUNTEER INFANTRY

Enlisted June 1, 1861, at Troy, New York, for 2 years, age 18.
Born on October 8, 1846, occupation: farmer.
5' 11" tall, blue eyes, auburn hair, light complexion.
Not stated June to 31 December 1861.
Shown as present January 1862 to June 1863.
Promoted to corporal March 1, 1862.
September 1862 on detached service at brigade headquarters.
Promoted to sergeant November 1, 1862.
Mustered out on June 8, 1863.
Mustered in August 26, 1863, at Saratoga Springs as a private in Company C, 2nd Regiment New York Veteran Cavalry for 3 years.
Remustered on September 1, 1863, as a 2nd lieutenant Company F.
Shown as present September 1863 to December 1863.
Promoted to 1st lieutenant on December 18, 1863.
Not stated January/February 1864.
Shown as present March/April 1864.
On leave of absence May/June 1864.
On July 31, 1864, on an illegal pass; captured.
Dismissed from service August 11, 1864.
Changed to honorable discharge February 26, 1865, when exchanged at Red River, Louisiana.
Died April 28, 1927.
Saw action at White Sulphur Springs, Gainesville, Groveton, and 2nd Bull Run, possibly at South Mountain, Antietam, Fredericksburg and Chancellorsville as a member of 30th New York State Volunteers.

Left: Obv. Steele. *Right:* Rev. Steele.

Private Edward Northrop

COMPANY A, 31ST NEW YORK STATE VOLUNTEER INFANTRY

Mustered in May 24, 1861, at New York, New York, for 2 years, age 23.
Not stated June to February 1862.
Note "Deserted June 2, 1861" with no other information.

Present April 1862 to March 1863.
Note "Stoppage of $5 from his pay by sentence of Regt Court-Martial" February 1, 1863.
Present to mustered out on June 4, 1863.
Saw action at West Point, Seven Days before Richmond, Gaines' Mill, White Oak Swamp, Malvern Hill, 2nd Bull Run, South Mountain, Antietam, Fredericksburg, Chancellorsville.

Left: Obv. Northrop. *Right:* Rev. Northrop.

Private Chauncey J. Clum

Company B, 33rd New York State Volunteer Infantry

Mustered in May 22, 1861, at Elmira, New York, for 2 years, age 24.
Not stated July/August 1861 to November/December 1861.
November/December 1861 owed sutler $4.00.
Present January/February 1862 to July/August 1862.
September 1862 absent, wounded at Antietam September 17, 1862; wounded in leg?
September/October 1862 "sick from wounds hospital unknown."
Muster out roll dated June 2, 1863, noted "Ab sick of wounds at hospital since 17 September 1862 place unknown"
April 11, 1891, pension document says "Died Sept 30, 1862, of wounds."

Left: Obv. Clum. Private Clum's identification disc was excavated near a hospital site at Antietam. *Right:* Rev. Clum.

Private Clum's identification disc displayed on his badly worn headstone, #157 in the Antietam National Cemetery.

Buried in Antietam National Cemetery, grave #157.
Saw action at Williamsburg, Mechanicsville, Seven Days Battles, Antietam.

Private Robert Williams

COMPANY G, 33RD NEW YORK STATE VOLUNTEER INFANTRY

Mustered in May 23, 1861, at Elmira, New York, for 2 years, age 30.
Not stated July/August 1861 to November/December 1861.
Present January/February 1862 to muster out.
September/October 1862 pay deducted $2 by order of court-martial for "straggled."
Mustered out June 2, 1863, at Geneva, New York.
Owed government $1.19 for one cartridge box, belt, plate, one cap pouch and pick.
Saw action at Williamsburg, Mechanicsville, Seven Days Battles, Antietam, Chancellorsville.

Left: Obv. Williams. *Right:* Rev. Williams.

Sergeant John Johnson

COMPANY K, 34TH NEW YORK STATE VOLUNTEER INFANTRY

Enlisted on June 15, 1861, at Albany, New York, at age 24 as a corporal.
5' 6" tall, dark eyes, dark hair and dark complexion.
Carpenter at enlistment. Born in 1832?
Not stated June 1861 to April 1862.
Present May 1862 to June 1863.
Promoted to sergeant June 1, 1862.
Discharged June 30, 1863.

Postwar image of Sergeant Johnson (courtesy the Johnson Collection).

Died July 2, 1913.
Saw action at Yorktown, Seven Days Battles, Antietam, Fredericksburg.

Left: Obv. Johnson. *Right:* Rev. Johnson.

Private Hiram Hull

COMPANY I, 37TH NEW YORK STATE VOLUNTEER INFANTRY

Mustered in on October 9, 1861, at Elmira, New York, age 25, for 2 years.
Not stated September/October 1861 to January/February 1862.
Present March/April 1862.
Absent May/June 1862, sent to hospital June 2.
Wounded by spent ball or piece of shell at Fair Oaks.
Absent July/August 1862.
Present September/October 1862 to April 10, 1863.
Mustered out June 22, 1863.
Mustered in May 11, 1864 in Company D, 179th NYSV, age 27, for 3 years, was a farmer.
5' 9" tall, grey eyes, brown hair, fair complexion.
Present May/June 1864 to muster out.

Postwar image of Private Hull (courtesy U.S. Army Military History Institute).

Left: Obv. Hull. *Right:* Rev. Hull.

Promoted to corporal June 1864?
Promoted to sergeant April 1, 1865.
Discharged June 8, 1865.
Died March 31, 1908.
Saw action at Yorktown, Fair Oaks, 2nd Bull Run, Chantilly, Fredericksburg, Chancellorsville, Cold Harbor, Petersburg, The Crater, Weldon Railroad, Poplar Spring Church, Hatcher's Run, Fort Stedman.

Private Robert Stanton

Company A/I/D, 43rd New York State Volunteer Infantry

Mustered in on February 25, 1862, at Albany, New York, age 23, for 3 years.
5' 5" tall, blue eyes, brown hair, light complexion, was a farmer.
Born Cooperstown, New York.
Present March/April 1862 to July/August 1862.
Claimed "At Fair Oaks part of a tree (due to enemy shell) fell on me and injured my (left) shoulder" (no medical record during war of this in his file).
July 18, 1862, transferred to Company D.
Appears on hospital muster roll, attached July 1, 1862, as cook.
Present on hospital roll at Franklin's Corps Hospital, Hagerstown.
September/October 1862 present.
Not stated November/December 1862, 6th Corps hospital, Hagerstown.
Present January/February 1863 to July/August 1863, shown as nurse/cook.
September/October 1863 note "entitled to $32.50 for services rendered at Semanary(?) Hospital at Hagerstown."
Note September 1, 1863, "last date clothing account settled owed Wm. H. Gomersall(?) sutler $10."
Present September/October 1863 to May/June 1864 at Hagerstown Hospital.
Present July/August 1864 at City Point, Virginia, attendant at division hospital.
Mustered out of 43rd Regiment due to consolidation into 43rd Battalion October 11, 1864.
November/December 1864 to discharge February 25, 1865, present at hospital.
Last paid on October 31, 1864, and due $100 bounty at discharge.
Received $5.97 for clothing allowance.
Died July 25, 1918.
Saw action at Yorktown, Fair Oaks?

Left: Obv. Stanton. *Right:* Rev. Stanton.

Private William Graham
Company K, 56th New York State Volunteer Infantry
Mustered in on March 29, 1865, in New York City, New York, at age 19.
Mustered out October 17, 1865, at Charleston, South Carolina.

Left: Obv. Graham. **Right:** Rev. Graham.

Sergeant Joseph H. Toms
Company H, 57th New York State Volunteer Infantry
Mustered in on September 17, 1861, age 25, for 3 years, at New York.
5′ 7″ tall, dark eyes, dark hair and dark complexion, was a painter.
Not stated October 1861 to December 31, 1861.
Present January/February 1862.
Not stated March/April 1862.
Present May/June 1862 to July/August 1862.
Detached to hospital department August 12, 1862.
Rest of service assigned to ambulance corps.
Mustered out October 15, 1864, near Petersburg, Virginia.
Due soldier $31.17 for clothing and $100 bounty.

Left: Obv. Toms. **Right:** Rev. Toms.

Died May 18, 1890.

Saw action at Yorktown, Fair Oaks, Gaines' Mill, Peach Orchard, Savage's Station, White Oak Swamp, Malvern Hill, Antietam, Fredericksburg, Chancellorsville, Gettysburg, Bristoe Station, Mine Run, Wilderness, Po River, Spotsylvania, North Anna, Tolopotomoy, Cold Harbor, Petersburg.

Private William C. Taylor

COMPANY H, 57TH NEW YORK STATE VOLUNTEER INFANTRY

Mustered in on September 17, 1861, age 28, for 3 years, at New York.

5' 5" tall, blue eyes, light hair and light complexion.

Born in New Jersey, was a blacksmith.

Not stated October 1861 to January/February 1862.

Record missing for March/April.

Present May/June 1862 to October 1864.

Due U.S. 1 Enfield rifle, 1 bayonet and sheath, 1 set belts and boxes.

July 1862 assigned as cook in regimental hospital.

Remained assigned to regimental hospital to muster out.

Mustered out October 15, 1864, near Petersburg, Virginia.

Due U.S. $6.90 for clothing and due soldier $100 bounty.

Last paid February 29, 1864.

Died January 14, 1913.

Saw action at Yorktown, Fair Oaks, Gaines' Mill, Peach Orchard, Savage's Station, White Oak Swamp, Malvern Hill, Antietam, Fredericksburg, Chancellorsville, Gettysburg, Bristoe Station, Mine Run, Wilderness, Po River, Spotsylvania, North Anna, Tolopotomoy, Cold Harbor, Petersburg.

Left: Obv. Taylor. Private Taylor's identification disc was excavated at his regiment's 1863–1864 winter camp in Virginia. *Right:* Rev. Taylor.

Private Robert H. Hurst

COMPANY F, 61ST NEW YORK STATE VOLUNTEER INFANTRY

Enlisted on September 17, 1861, at New York, New York, age 19.

5' 4½" tall, blue eyes, light hair, light complexion, born in 1832?

Not stated October 1861 to December 1861.

Transferred to Company F, November 1, 1861.
Present January 1862 to June 1862.
Sick in hospital at Alexandria, Virginia.
July/August moved to hospital in Providence, Rhode Island.
Died at hospital September 22, 1862.
Saw action at Yorktown, Seven Days Battles.

Left: Obv. Hurst. *Right:* Rev. Hurst.

Private Albert E. Wilmont

Company A, 61st New York State Volunteer Infantry

Enlisted on October 18, 1861, at New York, New York, as a corporal, age 18, born in 1838?
Not stated October 1861 to December 1861.
Reduced to private November 24, 1861.
Present January 1862 to April 1862.
Sent to hospital April 20, 1862, absent to June 1862.
Present July to November 1862.
Transferred to Company F September 20, 1862.
Sent to Finly Hospital, Washington, D.C., November 16, 1862.
Discharged for disability at Washington, D.C., on December 17, 1862.

Left: Obv. Wilmont. *Right:* Rev. Wilmont.

Hurt back building cordoroy roads at Yorktown.
Died December 21, 1899.
Saw action at Yorktown, Antietam?

Private Peter H. Flick

COMPANY D, 70TH NEW YORK STATE VOLUNTEER INFANTRY (EXCELSIOR BRIGADE)

Resided in Lancaster, Pennsylvania.
Mustered in on July 10, 1861.
Drowned in the Potomac River on March 17, 1862.

Left: Obv. Flick. *Right:* Rev. Flick.

Private L. D. Finch

COMPANY E, 72ND NEW YORK STATE VOLUNTEER INFANTRY

Enlisted on July 21, 1861, at Staten Island, New York, for 3 years, as a private.
Age 23, worked as a laborer, 5' 6½" tall, blue eyes, light hair.
Not stated July 1861 to November 1861.
Sick in quarters December 1861.
Present from January 1862 to February 1862.

Obv. Finch. Rev. Finch.

March 1862 3rd Regiment redesignated 72rd Regiment.
Present March 1862 to November 1862.
Reported wounded, Malvern Hill, July 1, 1862.
December 1862 detached to Captain James E. Smith's Fourth Independent New York Battery.
Shown as with Smith's Battery the rest of his service.
Sent to hospital, Fairfax Seminary, June 14, 1863.
Rest of service in hospital.
Mustered out June 19, 1864.
Died March 22, 1909.
Saw action at Williamsburg, Fair Oaks, Seven Days, Malvern Hill, 2nd Bull Run, Fredericksburg, Chancellorsville.

Private Samuel. H. Cotton

COMPANY I, 74TH NEW YORK STATE VOLUNTEER INFANTRY

Enlisted on October 6, 1861, at Dansville, New York, for 3 years as a private, age 21.
Appointed sergeant on October 7, 1861.
Not stated October 1861 to December 1861.
Listed as present from January 1862 to August 1862.
March 1862 5rd Regiment redesignated 74th Regiment.
Wounded in action at Bristoe Station, August 27, 1862.
Admitted to hospital on September 1, 1862.
In Eckington Hospital in Washington, D.C.
Died of wound on September 7, 1862.
Saw action at Williamsburg, Fair Oaks, Seven Days, Bristoe Station.

Left: Obv. Cotton. *Right:* Rev. Cotton.

Private John Fritzmaurice

COMPANY I, 74TH NEW YORK STATE VOLUNTEER INFANTRY

Enlisted on October 7, 1861, at New York, New York, in Company K for 3 years as a private.
Age 17, 5' 6" tall, dark complexion.
Blue eyes, dark hair, born in Cork, Ireland, a laborer.
Mustered in on December 8, 1861 (note age given is 20).
Not stated December 1861.

Listed as present from January 1862 to June 1862.
Transferred to Company I, January 1, 1862.
March 1862 5rd Regiment redesignated 74th Regiment.
Listed as absent without leave on August 28 roll.
Left at Alexandria(?)
Present September 1862/October 1863.
No record for November/December 1862.
In July/August 1863 sentenced to stoppage of $5.00 by general court-martial (S.O. #65 Headquarters 2nd Division 3rd Corps).
Absent November 1863–March 1864.
November 19, 1863, ordered to confinement with forfeiture of $8.00 per month for 6 months by general order (# 80 Headquarters 2nd Division 3rd Corps).
Returned to duty on April 26, 1864.
Transferred to Company H, 40th New York State Volunteers, on July 27, 1864.
40th New York State Volunteer regimental descriptive book says.
Age 28, 5' 4" tall, light complexion, grey eyes, light hair.
Born in Greenwich, Scotland [doesn't sound like same person — authors]
Discharged on October 21, 1864, near Petersburg, Virginia.
Saw action at Williamsburg, Fair Oaks, Seven Days (?), Bristoe Station, Groveton, 2nd Bull Run, Fredericksburg, Chancellorsville, Gettysburg, Wilderness, Spotsylvania, Spotsylvania Court House, Bloody Angle, North Anna, Cold Harbor, Petersburg.

Obv. Fritzmaurice. Private Fritzmaurice's identification disc was damaged during excavation at an unknown time and location. When the ribbon was added is also unknown. Rev. Fritzmaurice.

Private George Moore Jr.

COMPANY A, 76TH NEW YORK STATE VOLUNTEER INFANTRY

Enlisted on October 8, 1861, at Virgil, New York, for 3 years, as a private, age 21.
Not stated October 1861 to December 1861.
Present January 1862 to April 1862.
May/June 1862 not stated.
Present July/August 1862.
Sent to hospital September 19, 1862, absent to May 1863.
May/June 1863 assigned to Invalid Corps.
Present December 8, 1863, to April 1864.

Promoted to sergeant April 1, 1864.
Assigned to Brigade Band May/June 1864.
Present July/August 1864.
Assigned to brigade headquarters September 1864.
Discharged on October 10, 1864.
Died December 31, 1904.
Saw action at 2nd Bull Run, South Mountain, Antietam?, Wilderness, Spotsylvania, Cold Harbor, Petersburg.

Left: Obv. Moore. *Right:* Rev. Moore.

Private Thomas B. Aumack

Company C, 87th New York State Volunteer Infantry

Mustered in on November 5, 1861, at Brooklyn, New York, for 3 years.
Age 18, 5' 8" tall, blue eyes, light hair, light complexion.
Not stated November 1861 to February 1862.
Listed as present March/April 1862.
Absent May 1862, sick.
Present May/June 1862.

Obv. Aumack. Rev. Aumack. Private Aumack had his home town of "RARITAN. N.J." stamped on his identification disc even though he served in a New York regiment.

September 6, 1862, consolidated with 40th New York State Volunteers.
Discharged September 23, 1862, at Fort Monroe.
Died at home on October 13(?), 1862.
Saw action at Williamsburg, Fair Oaks, Seven Days, Oak Grove.

Private George A. Matthews

COMPANY I, 87TH NEW YORK STATE VOLUNTEER INFANTRY

Mustered in on November 15, 1861, for 3 years, as a private.
Age 27, 5' 7½" tall, blue eyes, light hair, light complexion.
Born Longford, Ireland.
Not stated November 1861 to February 1862.
Present March/June 1862.
Not stated July/August 1862.
No records August to October 1862.
Transferred to Company I, 40th New York State Volunteers.
87th New York State Volunteers disbanded September 6, 1862.
Present November/December 1862.
Captured December 13, 1862, Fredericksburg.
Confined in Richmond December 17, 1862, paroled City Point, January 9, 1863.
Listed as deserted December 22, 1862.
Returned to duty May 15, 1863.
Wounded in ankle (slight) July 2, 1863, at Gettysburg.
Absent July to December 1863, sick in Alexandria Hospital.
Not stated January/February 1864.
Present February to April 1864 returned to Company F from hospital.
Present March to May 1864.
May 1864 returned to hospital.
Present July 1864 to April 1865.
Promoted to corporal October 1864.
Promoted to sergeant March 1, 1865.
Promoted to 1st sergeant May 14, 1865.
Mustered out June 27, 1865.
Saw action at Williamsburg, Fair Oaks, Seven Days, Oak Grove, Fredericksburg, Gettysburg,

Left: Obv. Matthews. *Right:* Rev. Matthews.

Wilderness, Spotsylvania, Cold Harbor, Weldon Railroad, Deep Bottom, Poplar Spring Church, Boydton Road, Hatcher's Run, Fort Stedman, White Oak and Petersburg.

1st Lieutenant John G. Mitchill
Company D, 101st New York State Volunteer Infantry

Mustered into Company A, 71st New York State Volunteers, on May 3, 1861, for three months. Had blue eyes and was 5' 9" tall.
Mustered out on July 30, 1861.
Mustered into the 101st as a 1st lieutenant on November 1, 1861, for three years.
Received a medical discharge for sciatica on September 9, 1862.
Saw action at Oak Grove, White Oak Swamp, Savage's Station, Glendale and Malvern Hill.

Left: Obv. Mitchill. **Right:** Rev. Mitchill.

Private George Ridley
Company D, 145th New York State Volunteer Infantry

Mustered in on September 11, 1862, at New York, New York, for 3 years, age 19.
5' 4" blue eyes, fair complexion, brown hair.
Present October 1862.

Left: Obv. Ridley. **Right:** Rev. Ridley.

Absent November 1862.
Detailed as Brigade Teamster November 18, 1862.
Present January/February 1863 as teamster.
Present as a patient at Stone U.S.A. Hospital on February 20, 1863.
Discharged on February 26, 1863, at General Hospital, Washington, D.C., for disability (going blind in right eye, left eye blind from infancy, according to Certificate of Disability for Discharge).
Saw no action.

Private Joseph Lewis

Company I, 147th New York State Volunteer Infantry

Mustered in on July 21, 1863 at Elmira, New York, for 3 years.
Age 18, 5' 8", blue eyes, dark complexion, brown hair, occupation listed as farmer.
Present September/October 1863.
Absent November/December on detached service at corps headquarters.
Present January/February 1864 to June 3, 1864.
March/April 1864 due government $0.52 for one cartridge box, plate and cap pouch.
Wounded at Bethesda Church.
Gun shot wound of left thigh, third bullet(?).
Absent July/August to November/December, sick at 3rd Division hospital 5th Corps.
Absent January/February 1865 in U.S. hospital Washington, D.C.
May have been granted a furlough.
Based on letter dated March 17, 1865, from a local doctor asking for an extension of his furlough due to expire on March 19, Private Lewis may have never returned from the furlough due to chronic diarrhea.
March 31, 1865 listed as deserter from Harewood Hospital.
Listed as absent, sick, on company muster out roll dated June 7, 1865.
Saw action at Mine Run, Wilderness, Spotsylvania, North Anna and Bethesda Church.

Left: Obv. Lewis. Private Lewis' identification disc was excavated in his regiment's 1863–1864 winter camp in Virginia. *Right:* Rev. Lewis.

Private James H. Pierson

COMPANY D, 153RD NEW YORK STATE VOLUNTEER INFANTRY

Mustered in on October 17, 1862, at Johnston, New York, for 3 years.
Age 18, 5' 4½", brown eyes, light complexion, brown hair, occupation listed as clerk/glove maker.
Born on May 29, 1844.
Absent, sick, at Johnston, New York, October 1862.
Present December 1862 to June 1864.
March–April 1864 stoppage of $19.25 for complete Springfield musket.
May/June 1864 stoppage of $1.00 for error in musket payment of $19.75.
Absent without permission on June 10, 1864.
Missed tattoo roll call, pleaded guilty, fined $1.00.
July/August 1864 absent, sick, for chronic diarrhea from Red River Expedition.
Present September/October 1864.
Detached service with ambulance corps at Camp Sheridan,
November 1864 to February 1865.
Present March/April 1865.
Absent, sick, at Finley Hospital, Washington, D.C.
May 20, 1865 on march to Savannah received sunstroke.
May/June 1865 mustered out at hospital by telegram on July 17, 1865.
Noted on pension form dated April 7, 1915, "old bachelor."
Also noted in file "Subject to periodical sprees lasting 1 to 5 days."
Died on November 16, 1929.
Saw action at Red River, Sabine Cross Roads, Fort Stevens, Opequon, Fisher's Hill and Cedar Creek.

Left: Obv. Pierson. Private Pierson probably had this identification disc made as a keepsake for his mother based on the presentation "TO. MY. MOTHER." *Right:* Rev. Pierson.

Private Orrin Holmes

COMPANY A, 157TH NEW YORK STATE VOLUNTEER INFANTRY

Enlisted on August 11, 1862, at Hamilton, New York, at age 18.
5' 6" tall, blue eyes, light hair.
Present September 1862 to June 1863.
Wounded first day at Gettysburg.

In hospital July 1, 1863, to March 3, 1864.
Transferred to Veteran Reserve Corps.
Discharged on September 26, 1864.
Died January 21, 1919.
Saw action at Chancellorsville and Gettysburg.

Left: Obv. Holmes. Private Holmes' identification disc is unusual in that the small tassel is very rarely found with a disc. *Right:* Rev. Holmes. Who D. Perry was has not been discovered.

Corporal Patrick O'Neal

Company B, 2nd New York Heavy Artillery

Mustered in on September 18, 1861.
Promoted to corporal on February 23, 1864.
Enlisted as a Veteran Volunteer on February 18, 1864.
Transferred to Veteran Reserve Corps on April 2, 1865.
Discharged on August 12, 1865.
Saw action during the Overland Campaign and Petersburg.

Left: Obv. O'Neal. *Right:* Rev. O'Neal.

Private Henry A. Lown
Company H, 4th New York Heavy Artillery

Enlisted on January 23, 1862, at New York, New York, age 29, born in Rheinbeck, New York.
5' 10", complexion light, eyes blue, hair brown, occupation listed as farmer.
Discharged of disability at Fort Marcy on December 29, 1862.
Re-enlisted on December 24, 1863, in Company H, 16th Regiment New York Artillery.
Furloughed on November 1, 1864.
Never returned to duty.
Listed as a deserter on November 30, 1864.
Died while on furlough, December 17, 1864, at home.
Saw no action.

Left: Obv. Lown. **Right:** Rev. Lown.

Ohio

Private James McInnes
Company H, 5th Ohio Infantry

Mustered in June 19, 1861, at Camp Dennison, Ohio, for 3 years.
Age 21, 5' 2" tall, hazel eyes, light hair, light complexion.

Obv. McInnes. Rev. McInnes.

Born Glasgow, Scotland, was an iron molder.
Not stated June 1861 to February 28, 1862.
Present March 1862 to August 1862.
Absent September 1862 to December 1862, sick at Warrington Junction, July,
Later at Frederick, Maryland.
In Cliffbourne Hospital, Washington, D.C., on September 13 (17), 1862.
Discharged for disability on December 22 (29?), 1862, for chronic diarrhea.
Died April 13, 1892.
Saw action at First Winchester, Port Republic, and Cedar Mountain.

Private Owen Sullivan

COMPANY G, 61ST OHIO INFANTRY

Mustered in January 16, 1862, at Cincinnati, Ohio, for three years as a private.
Age 35, 5' 6" tall, grey eyes, black hair, dark complexion.
Born in Ireland, was a laborer.
Present from May 1862 to April 1863.
May/June 1863 not stated.
Reported becoming overheated in Battle of Chancellorsville,
May 1, 2, 3, 1863, leading to blindness in 1873.
July/August 1863 absent in hospital.
Discharged September 12, 1863.
Owen died in the summer of 1891 based on an entry in his file. Last paid June 4, 1891. dropped "due to death."
Saw action at Freeman's Ford, Second Bull Run, and Chancellorsville.

Obv. Sullivan. Rev. Sullivan. On Private Sullivan's identification disc the sutler stamped the wrong enlistment date of 1861 instead of 1862.

Private John Wright

COMPANY H, 61ST OHIO INFANTRY

Mustered in February 4, 1862, at Bellaire, Ohio, for three years as a private.
Promoted to sergeant on April 24, 1862.
Present from April 1862 to October 1862.
November/December listed as an "orderly."

Promoted to first sergeant on October 31, 1862.
Sent to a hospital on July 3, 1863.
Wounded at the Battle of Gettysburg on July 2, 1863 ("gunshot wound in the left breast").
In Cuyler U.S.A. General Hospital at Germantown near Philadelphia, Pennsylvania, on July 6, 1863.
January 28, 1864, transferred to the 1st Battalion Invalid Corps.
July 22, 1864, received a presidential appointment as a 1st lieutenant in Company B of the 119th Regiment of Colored Troops.
Enlisted for 3 years, mustered in on February 3, 1865.
Shown as "present" May 1865 to August 1865.
July to November 1865 commanding Company H.
December 13, 1865, promoted to captain, commanding Company K.
Mustered out on April 27, 1866.
Wright died in the fall of 1905 or spring of 1906 based on an entry in his file dated June 10, 1906, that he was dropped because of "death date not given."
Saw action at Freeman's Ford, Second Bull Run, Chancellorsville, Gettysburg.

Left: Obv. Wright. *Right:* Rev. Wright. Sergeant Wright left his hometown of "GERMANTOWN. PENN." to serve in an Ohio regiment.

Pennsylvania

Private George W. Morgan

Company K, 3rd Pennsylvania Reserve Volunteer Corps[1]

Mustered in on May 27, 1861, at Philadelphia, Pennsylvania, for 3 years, at age 19.
Had blonde hair and hazel eyes.
Wounded in action at Fredericksburg on December 13, 1862, and reported as deserted.
Returned and reinstated under presidential proclamation on March 7, 1863.
Present until mustered out on June 17, 1864.
Saw action in the Peninsular Campaign, 2nd Manassas, South Mountain, Antietam, Fredericksburg, and Cloyd Mountain.

Left: Obv. Morgan. *Right:* Rev. Morgan.

Sergeant Isaac E. Lewis
Company H, 3rd Pennsylvania Reserve Volunteer Corps

Mustered in on June 18, 1861, for three years at the age of 32.
He was self-employed as a coach trimmer.
Wounded at 2nd Bull Run on August 30, 1862, by gunshot that fractured his thigh.
Died from that wound on September 24, 1862, at the U.S. General Hospital at Fairfax Seminary, Virginia.
Saw action on the Peninsular Campaign, 2nd Bull Run.

Left: Obv. Lewis. *Right:* Rev. Lewis.

Private James A. Morrison
Company D, 5th Pennsylvania Reserve Volunteer Corps

Enlisted June 5, 1861, mustered in on June 12 at Harrisburg, Pennsylvania, for 3 years.
At age 22, 5' 7½" tall, gray eyes, brown hair, dark complexion.
Born Lancaster, Pennsylvania, was a machinist.
Not stated June 1861 to December 1861.
Present January 1862 to June 1862.
Wounded in leg and missing June 30, 1862, at Savage Station/Gaines' Mill.

Absent July 1862 to October 1862, in hospital at Fort Monroe, August 5.
Paroled at City Point, August 3, 1862.
September in Chesapeake Hospital.
Present November 1862 to December 1863.
Re-enlisted Veteran Volunteers December 22, 1863.
Present December 1863 to June 1864.
June 6, 1864, transferred to Company D, 191st Pennsylvania.
Wounded in August 1864 (not confirmed in records).
Present July 1864 to January/February 1865.
Absent March 1865.
POW March 31, 1865, at Boydtown Plank Road.
Paroled at Aiken's Landing, April 2, 1865.
Reported to camp in Maryland, April 4, 1865.
Reported to Camp Parole, Maryland, April 6, 1865.
Mustered out at Annapolis, Maryland, May 29, 1865.
Saw action at Mechanicsville, Gaines' Mill, Fredericksburg, Gettysburg, Bristoe Station, Rappahannock Station, Mine Run, Wilderness, Cold Harbor and Petersburg.

This ambrotype was purchased with Morrison's identification disc and therefore it is believed to be an image of Private Morrison.

Left: Obv. Morrison. *Right:* Rev. Morrison.

Private Julian A. (H) Bradley

COMPANY H, 5TH PENNSYLVANIA RESERVE VOLUNTEER CORPS

Mustered in on June 21, 1861, for three years.
Transferred to 191st Pennsylvania Volunteer Infantry on June 6, 1864.
Mustered out on June 28, 1865.
Saw action in the Peninsular Campaign, 2nd Manassas, South Mountain, Antietam, Fredericksburg, Gettysburg, Wilderness, Spotsylvania, Cold Harbor, Petersburg, Weldon Railroad, Five Forks, and Appomattox Court House.

Left: Obv. Bradley. *Right:* Rev. Bradley.

Private Samuel E. McCleary (McClary)
COMPANY C, 11TH PENNSYLVANIA RESERVE VOLUNTEER CORPS

Mustered in on July 29, 1861, at Washington, D.C., for 3 years at age 20.
Residence Sunbury, Pennsylvania.
Wounded in action at Gaines' Mill, Virginia, and captured in June 1862.
Paroled July 19, 1862.
Captured again at the Wilderness, May 5, 1864.
Paroled on February 27, 1865.
Saw action at Gaines' Mill, Gettysburg (Wheatfield) and the Wilderness.

Left: Obv. McCleary. *Right:* Rev. McCleary.

Private David D.P. Alexander
COMPANY G, 11TH PENNSYLVANIA RESERVE VOLUNTEER CORPS

Enlisted on July 5, 1861, at Armstrong County, Pennsylvania, age 24, as a musician.
5' 3" tall, blue eyes, light complexion, brown hair.
Born March 4, 1837, was a harness maker.
Present from August 1861 to December 1861.
Hospital duty December 1861 to June 1862.

Detached duty July and August 1862.
Present September 1862 until mustered out June 13, 1864.
Married on June 13, 1865, to Abigail May. They had ten children; seven survived.
He died on January 19, 1920, at Ingram, Pennsylvania.
Saw action at South Mountain, Antietam, Fredericksburg, Gettysburg, Bristoe Station, Rappahannock Station, Mine Run, Wilderness, and Bethesda Church.

Left: Obv. Alexander. *Right:* Rev. Alexander.

Private James H. Coffin
Company E, 12th Pennsylvania Reserve Volunteer Corps

Mustered in on August 10, 1861, at Harrisburg, Pennsylvania, for 3 years, at age 20.
Present for duty from muster in until mustered out on June 11, 1864.
Saw action in the Peninsular Campaign, 2nd Manassas, South Mountain, Antietam, Fredericksburg, Gettysburg, Wilderness and Spotsylvania.

Left: Obv. Coffin. *Right:* Rev. Coffin.

Private Thomas R. Senior
Company K, 26th Pennsylvania Volunteer Infantry

Mustered in on June 6, 1861, at Philadelphia, Pennsylvania, for 3 years at age 20.
He was 5' 8" tall and was employed as a farmer.

Received a gunshot wound to right leg May 3, 1862, at Williamsburg, Virginia.
Received a gunshot wound in left arm July 1, 1862, at Charles City Cross Road, Virginia (Malvern Hill).
Admitted to the hospital on July 4, 1862.
Incorrectly listed as deserter while in the hospital.
Transferred to the Veteran Reserve Corps on December 3, 1863.
Medical discharge received on July 24, 1865.
Charge of desertion removed on July 22, 1871.
Saw action in the Peninsular Campaign.

Left: Obv. Senior. *Right:* Rev. Senior.

Private John Shaffer

Company F, 45th Pennsylvania Volunteer Infantry

Mustered in on October 16, 1861, at Harrisburg, Pennsylvania, for three years at age 18.
He was 5' 5" tall, with blue eyes, blonde hair, was born in Andes, New York, and employed as a laborer.
In hospital November/December 1861.
Present for duty from January to September 1, 1862, when he was sent to the hospital.
Present for duty November 25, 1862, until August 24, 1863.

Left: Obv. Shaffer. Rev. Shaffer.

Transferred for duty as a teamster.
Lost musket and accoutrements at Harpers Ferry and charged $16.46.
Enlisted as a Veteran Volunteer on February 2, 1864.
Wounded in action at Petersburg on June 18, 1864.
Mustered out on July 17, 1865.
Saw action at South Mountain, Jackson, Miss., Spotsylvania, Cold Harbor and Petersburg.

Private Henry T. Rice

COMPANY G, 45TH PENNSYLVANIA VOLUNTEER INFANTRY

Mustered in on September 18, 1861, at the age of 20.
He was 6' tall, had gray eyes, and was employed as a farmer.
Received a flesh wound in breast at the Wilderness on May 6, 1864.
Wounded in left hand at Cold Harbor on June 3, 1864.
Transferred to Veteran Reserve Corps on January 1, 1865.
Saw action at South Mountain, Jackson, Mississippi, Spotsylvania and Cold Harbor.

Left: Obv. Rice. *Right:* Rev. Rice.

Private Jacob P. George

COMPANY B, 47TH PENNSYLVANIA VOLUNTEER INFANTRY

Mustered in on August 30, 1861, at Allentown, Pennsylvania, for the duration.
Present from muster in until June 1864.
Re-enlisted as a Veteran Volunteer on October 10, 1863.
Sick at Harpers Ferry during July and August 1864.
Present until muster out on December 25, 1865.
Saw action at Antietam, Chancellorsville, Sabine Cross, Louisiana, Opequon, Fisher's Hill, and Cedar Creek.

Corporal Milton H. Dunlap

COMPANY H, 47TH PENNSYLVANIA VOLUNTEER INFANTRY

Mustered into 128th Pennsylvania Volunteer Infantry at Harrisburg, Pennsylvania, for nine months: on August 1862, as a private at the age of 21.
He was 5' 4" tall and had green eyes and brown hair.

Left: Obv. George. *Right:* Rev. George.

Mustered out in May 1863.
Mustered into 47th Pennsylvania Volunteers, on December 12, 1863, at Newport, Pennsylvania, for 3 years.
Present January 1864 until June 1864.
In hospital from July 1864 until October 1864.
Present November 1864 until June 1865.
Saw action at Antietam, Chancellorsville, Sabine Cross, Louisiana, and Cedar Creek.

Left: Obv. Dunlap. *Right:* Rev. Dunlap.

1st Sergeant John W. Jenkins

Company F, 48th Pennsylvania Volunteer Infantry

Mustered in on August 22, 1861, at Minersville, Pennsylvania, for 3 years at age 20.
He was 5' 7" tall, had hazel eyes, dark hair, and was employed as a miner.
He was wounded at Antietam, Maryland, by a shell fragment which caused a fractured skull.
A medical discharge was granted on December 4, 1862.
Saw action at 2nd Manassas and Antietam.

Left: Obv. Jenkins. *Right:* Rev. Jenkins.

Private David Bossart

Company I, 49th Pennsylvania Volunteer Infantry

Enlisted on October 24, 1861, at Juniata County, Pennsylvania, for 3 years, at age 24.

Present from July 1862.

Transferred to Company A on January 1, 1863.

Discharged on October 23, 1864.

Died August 31, 1887.

Saw action at Crampton's Gap, Salem Church, Gettysburg, Rappahannock Station, Mine Run, Wilderness, Laurel Hill, Spotsylvania, Cold Harbor, and Opequon.

Postwar view of Private Bossart wearing his identification disc below his 6th Corps Badge (courtesy U.S. Army Military History Institute).

Left: Obv. Bossart. *Right:* Rev. Bossart.

Private Andrew Lucas

Company A, 49th Pennsylvania Volunteer Infantry

Mustered in on August 31, 1861, at Milesburg, Pennsylvania, for 3 years, at age 21.
Present May/June 1862.
Wounded in action June 27, 1862, at Golding's Farm or Gaines' Mill.
July/August present at General Hospital at Baltimore, Maryland, November/December 1862 discharged at Liberty Park, Baltimore, October 31 on surgeon's certificate of disability.
Stiff elbow of left arm due to gunshot wound.
Mustered in to Company E. 184th Pennsylvania Volunteer Infantry.
For 3 years was a brakeman.
5' 9" tall, grey eyes, fair complexion, light hair.
Present May 12 to June 30.
Shown as corporal to May 1, 1865.
Reduced to private?
Mustered out on July 14, 1865.
Died before January 4, 1905, based on the stoppage of his pension, based on note in file.
Saw action at Yorktown, Williamsburg, Gaines' Mill, Cold Harbor, and Petersburg.

Left: Obv. Lucas. *Right:* Rev. Lucas.

Private Hall Henry (Henry Hall)

Company H, 53rd Pennsylvania Volunteer Infantry

Mustered in on October 23, 1861.
Probably captured shortly after Cold Harbor.
Died as POW at Florence, South Carolina.
Buried at Andersonville, Georgia.
Saw action at Gaines' Mill, Fair Oaks, Fredericksburg, Chancellorsville, Gettysburg, Wilderness, Spotsylvania, and Cold Harbor.

Private John I. Guigher

Company H, 56th Pennsylvania Volunteer Infantry

Mustered into the 56th Pennsylvania Volunteer Infantry on January, 25 1862, as a corporal.
At enlistment he was 19 years old, 5' 11" tall, had brown eyes and sandy hair, was employed as a "roller," and signed his enlistment by mark.

Left: Obv. Henry. *Right:* Rev. Henry.

Reduced to private on October 11, 1862.
Served as a division teamster from September 1863 until April 1864.
Enlisted as a Veteran Volunteer on February 12, 1864.
Mustered out on July 1, 1865.
Saw action at Gainesville (Brawner's Farm), Second Manassas, South Mountain, Antietam, Fredericksburg, Chancellorsville, Gettysburg, the Overland Campaign, Petersburg, Weldon Railroad, and Hatcher's Run.

Left: Obv. Guigher. *Right:* Rev. Guigher.

Private James McIntyre

Company A, 56th Pennsylvania Volunteer Infantry

Muster in on March 6, 1862, at Wayne County, Pennsylvania, for 3 years, as a private.
Age 19, 5' 5" tall, hazel eyes, dark hair, light complexion, was a farmer.
Present from March/April 1862 to May/June 1862.
July/August 1862 absent, sick, in hospital, in Philadelphia, Pennsylvania.
Absent June 30 to October 31.
May have returned to regiment.
Deserted on December 24, 1862 from near Pratt's Landing.
(Regiment was camped there on January 17, 1863.)
Saw action at Fredericksburg?

Left: Obv. McIntyre. *Right:* Rev. McIntyre.

Private Francis Reed
Company D, 56th Pennsylvania Volunteer Infantry

Mustered in January 18, 1862, for 3 years, at age 22.
He was 5' 4" tall, had chestnut hair and gray eyes, and was employed as a mariner.
Taken prisoner at Gettysburg July 1, 1863.
Enlisted as a Veteran Volunteer on February 20, 1864.
Wounded severely in thigh on May 8, 1864, at Spotsylvania.
Returned to duty November 24, 1864.
Received a shell wound to his right leg on November 30, 1864, at Petersburg.
Mustered out on July 25, 1865.
Saw action at Gainsville (Brawner's Farm), 2nd Manassas, South Mountain, Antietam, Fredericksburg, Chancellorsville, Gettysburg, Wilderness, Spotsylvania, and Petersburg.

Left: Obv. Reed. *Right:* Rev. Reed.

Private Christian Kerr (Carr)
Company K, 56th Pennsylvania Volunteer Infantry

Mustered in on April, 5 1862, at Montrose, Pennsylvania, at the age of 34.
He was 5' 6" tall, his occupation was a farmer, he was a native of Germany, and resided in Rockhill, Pennsylvania.

He was A.O.L. during May and June 1862.

Present for duty until re-enlisted as a Veteran Volunteer on March 29, 1864.

Wounded in action at Weldon Railroad, Virginia, on August 20, 1864, by gunshot in left of neck which exited on the right side of his neck.

Mustered out on April 5, 1865.

Saw action at Gainesville (Brawner's Farm), 2nd Manassas, Chancellorsville, Gettysburg, and the Overland Campaign.

Left: Obv. Carr. *Right:* Rev. Carr.

Corporal Reuben S. Kunkle

Company H, 67th Pennsylvania Volunteer Infantry

Enlisted January 12, 1862, in Monroe County, Pennsylvania.

Present from March 1862 until June 15, 1863.

Killed in action at Stephenson's Depot, Virginia (2nd Winchester).

Saw action at Berryville, Fredericksburg and 2nd Winchester.

Left: Obv. Kunkle. It is believed that Private Kunkle's identification disc was excavated near Stephenson's Depot, where he was killed in action. *Right:* Rev. Kunkle.

Private Theodore F. Stratton
Company K, 69th Pennsylvania Volunteer Infantry

Enlisted on August 22, 1861, at Philadelphia, Pennsylvania, for 3 years, as a private.
Age 18, 5' 6" tall, blue eyes, dark hair.
Present from March 1862 to August 16, 1862.
Absent in hospital until March 1863.
Present until May 1864, in hospital until August 28, 1864, promoted to sergeant September 1864.
Promoted to 2nd lieutenant November 1864.
Promoted to captain April 1865.
Mustered out July 1, 1865.
Saw action at Yorktown, Fair Oaks, Peach Orchard, Savage's Station, Charles City Court House, Malvern Hill, Thoroughfare Gap, Gettysburg, Mine Run, Hatcher's Run, Dabney's Mill and Petersburg.

Left: Obv. Stratton. *Right:* Rev. Stratton.

Private George Allendorf
Company A, 82nd Pennsylvania Volunteer Infantry

Enlisted as corporal in Company A 31st Pennsylvania (later 82nd), July 27, 1861 at Philadelphia, Pennsylvania, for three years.

Left: Obv. Allendorf. *Right:* Rev. Allendorf.

Age 35, 5' 7" tall, hazel eyes, sandy hair.
Present September 1862 to December 1863.
Demoted to private July 13, 1863.
Deserted February 22, 1864.
Returned March 2, 1864.
Present March 1864 until discharged in September 1864.
Enlisted Veteran Volunteers March 8, 1865.
Mustered out in March 1866.
Died February 7, 1889.
Saw action at Antietam, Fredericksburg, Chancellorsville, Gettysburg, Mine Run, Cold Harbor, Fort Stevens, and Summit Point.

Private William Stewart

Company H, 83rd Pennsylvania Volunteer Infantry

Mustered in on August 21, 1862, at Wakefield, Pennsylvania, at the age of 23.
He was 5' 6" tall, had blue eyes, and brown hair, was a native of Ireland, and was employed as a farmer.
Present for duty from muster in until May 8, 1864.
Gunshot wounds to left shoulder and right hip, and captured May 8, 1864, at Spotsylvania.
Paroled August 22, 1864.
Served duty at College Green Barracks until April 30, 1865.
Mustered out on June 12, 1865.
Saw action at Peninsular Campaign, Gaines' Mill, 2nd Manassas, Fredericksburg, Chancellorsville, Gettysburg (Little Round Top), Wilderness, and Spotsylvania.

Left: Obv. Stewart. *Right:* Rev. Stewart.

Corporal Joseph W. Allen

Company M, 100th Pennsylvania Volunteer Infantry

Mustered in on September 5, 1861, as a corporal.
At mustering in he was 19 years of age, 5' 10" tall, had gray eyes and brown hair, and was a farmer.
Promoted to sergeant on November 15, 1862.
Promoted to 1st sergeant on May 17, 1863.
Re-enlisted as a Veteran Volunteer on December 28, 1863.

Wounded and captured at Petersburg, Virginia, on July 30, 1864.
Escaped Danville P.O.W. prison on October 10, 1864.
Returned to Union lines in Tennessee on November 18, 1864.
Promoted to captain on January 12, 1865.
Under arrest in June 1865 pending filing of paperwork.
Mustered out on July 24, 1865.
Final reports submitted July 25, 1865.
Saw action at Secessionville, South Carolina, 2nd Manassas, Chantilly, Virginia, Fredericksburg, Vicksburg, Mississippi, Jackson, Mississippi, Knoxville, Tennessee, Overland Campaign, and Petersburg.

Left: Obv. Allen. *Right:* Rev. Allen.

Sergeant Reese J. Thomas

Company B, 102nd Pennsylvania Volunteer Infantry

Mustered into the 13th Pennsylvania Volunteer Infantry on April 25, 1861, for three months; mustered out on August 6, 1861; mustered into the 102nd Pennsylvania Volunteer Infantry on August 19, 1861, as a sergeant in Pittsburgh, Pennsylvania.
He was 24 years old, 5′ 5″ tall, and had been employed as a carpenter at muster in.
Absent on recruiting duty from July 1862 until February 1863.

Left: Obv. Thomas. *Right:* Rev. Thomas.

On December 19, 1863 he re-enlisted as a Veteran Volunteer.
He was wounded at Cold Harbor, Virginia.
Transferred to Veteran Reserve Corps on January 15, 1865.
Date of muster out unknown.
Saw action at Salem Church, Gettysburg, and in the Overland Campaign.

Private Daniel Colgan

Company I, 115th Pennsylvania Volunteer Infantry

Mustered into Company N of 1st Regiment California Infantry on July 1, 1861, at Washington, D.C., for 3 years as a private.
Listed as missing in action September/October 1861 at Balls Bluff; as of October 21, he owed the sutler $8.00.
Confined at Richmond October 24, 1861.
Regiment designation changed to Company H of the 71st Pa Infantry.
POW November/December 1861, paroled at Richmond February 19–20, 1862.
Discharged May 22, 1862, as a paroled POW at Washington, D.C.
Enlisted in Company I, 115th Pennsylvania, June 4, 1862; mustered in July 5, 1862, at Philadelphia for three years as a corporal.
Company descriptive book: Age 21, 5' 4" tall, fair complexion, blue eyes, light hair, born in Pantucket, Massachusetts, listed as laborer.
Present June 1862 to August 1863.
Promoted to sergeant August 1, 1863.
Present as sergeant to April 1864.
No muster out roll of Company I, 115th Pennsylvania.
Transferred to Company F, 110th Pennsylvania, June 1864.
Present July/August 1864 to April 1865.
Captured at Hatchers Run, March 25, 1865.
In Richmond on March 27 as POW.
Paroled at Boulware & A Wharf March 30, later at Camp Parole Annapolis, Maryland.
Reported as sick as of April 17, 1865.
Mustered out May 31, 1865, at Washington, D.C.
Died July 7, 1926.

Left: Obv. Colgan. Private Colgan's identification disc that was excavated in the regimental winter camp of 1863-1864 in Virginia. *Right:* Rev. Colgan.

Saw action at Ball's Bluff, Malvern Hill, Second Bull Run, Fredericksburg, Chancellorsville, Gettysburg, Mine Run, Wilderness, Cold Harbor(?), Petersburg, Deep Bottom, Poplar Spring Church, and Boydton Road/Hatcher's Run.

Private Jesse Henry

Company A, 119th Pennsylvania Volunteer Infantry

Mustered in August 19, 1862, at Philadelphia, Pennsylvania, for 3 years, at age 30.
He had a dark complexion, brown eyes, dark hair, and his occupation was an engineer.
He was present for duty until killed in action on November 7, 1863 at Rappahannock Station, Virginia.
Saw action at Fredericksburg, Salem Church, Gettysburg and Rappahannock Station.

Left: Obv. Henry. Author Larry B. Maier's great-great grandfather John E. Faust was also a member of the 119th Pennsylvania Volunteer Infantry when he was mortally wounded at the battle of Salem Church during the Chancellorsville Campaign in 1863. *Right:* Rev. Henry.

Corporal William E. Cooper

Company C, 119th Pennsylvania Volunteer Infantry

Mustered in on August 19, 1862, in Philadelphia, Pennsylvania, for three years.
At the time of mustering in he was 20 years old, 5' 6" tall, had blond hair and had been employed as a machinist.

Left: Obv. Cooper. *Right:* Rev. Cooper.

He was promoted to corporal on November 1, 1863.

He was absent, sick, in the hospital from May to July 1864. Company records show him as present but was killed in action by a wound to the chest on September 19, 1864, at Opequon.

He saw action at Fredericksburg, Salem Church, Gettysburg, Rappahannock Station and Opequon.

Private John B. Cassady (Musician)

COMPANY I, 119TH PENNSYLVANIA VOLUNTEER INFANTRY

Mustered in August 14, 1862, Philadelphia, Pennsylvania, for three years at age 19. Musician.
He was 5' 6" tall, and had a dark complexion, blue eyes, and blonde hair.
Present for duty from muster in until November 1864 when placed in the hospital.
Mustered out from the hospital on June 19, 1865.
Saw action at Fredericksburg, Salem Church, Rappahannock Station, and the Overland Campaign.

Left: Obv. Cassady. **Right:** Rev. Cassady.

Private Andrew J. Bleaderheiser

COMPANY H, 121ST PENNSYLVANIA VOLUNTEER INFANTRY

Muster in August 21, 1862, Philadelphia, Pennsylvania, as a private at age 26.
5' 5" tall, brown eyes, brown hair, occupation paper hanger.

Left: Obv. Bleaderheiser. **Right:** Rev. Bleaderheiser.

Present from muster in until August 14, 1864.
Died in hospital November 19, 1864, at Philadelphia from dropsy of the chest.
Saw action at Fredericksburg, Chancellorsville, Gettysburg, and during the Overland Campaign.

Private William J. Sheehan
COMPANY G, 121ST PENNSYLVANIA VOLUNTEER INFANTRY

Mustered in on August 19, 1862, at Philadelphia, Pennsylvania, for three years as a private.
Age 19, 5' 11½" tall, blue eyes, dark hair, listed as a painter.
Present from September 1862 to June 29, 1863.
Captured at Gettysburg July 1, 1863.
Paroled August 6, 1863.
Deserted September 19, 1863, from parole camp.
Saw action at Fredericksburg, Chancellorsville and Gettysburg.

Left: Obv. Sheehan. Private Sheehan's identification disc was purported to have been dug in a school yard in Buffalo, Kansas. *Right:* Rev. Sheehan.

Private William H. McCreery
COMPANY D, 135TH PENNSYLVANIA VOLUNTEER INFANTRY

Mustered in August 19, 1862, at Harrisburg, Pennsylvania, for nine months at age 23.
He was 5' 8" tall, had blue eyes, blonde hair, and was employed as a teacher.
Appointed company clerk on August 31, 1862, regimental postmaster on December 1, 1862, and brigade clerk in April 1863.
Discharged May 24, 1863.
Enlisted in the U.S. Signal Corps on February 6, 1864.
Served in Washington, D.C.
Mustered out on August 24, 1865.

Private Silas E. Elmendorf
COMPANY A, 137TH PENNSYLVANIA VOLUNTEER INFANTRY

Mustered in on August 6, 1862, at Canaan, Pennsylvania, for nine months.
Present for duty from August 1862 until mustered out on June 2, 1863.
Saw action at Antietam and Chancellorsville.

Left: Obv. McCreery. *Right:* Rev. McCreery.

Left: Obv. Elmendorf. *Right:* Rev. Elmendorf.

Sergeant John F. Swiler

Company C, 148th Pennsylvania Volunteer Infantry

Mustered in on August 27, 1862, at Centre County, Pennsylvania, for three years.
Present from muster in until May 3, 1863.
Wounded in action at Chancellorsville on that date.
Returned to duty September 9, 1863.
Present for duty until taken prisoner at Petersburg on June 22, 1864.
Died while a prisoner of war.
Saw action at Chancellorsville, during the Overland Campaign, and at Jerusalem Plank Road, Virginia.

Private Andrew L. Whitehill

Company C, 148th Pennsylvania Volunteer Infantry

Enlisted on August 19, 1862, at Centre County, Pennsylvania, for 3 years as a private.
Age 22, 6' tall, blue eyes, dark hair.
Present from September 1862 to May 3, 1863.
Wounded in right thigh May 3, 1863, at Chancellorsville.

Left: Obv. Swiler. The heavy wear on this disc was purportedly due to an owner carrying it in his wallet for years as a good-luck charm. *Right:* Rev. Swiler.

Sent to hospital then transferred to Veteran Reserve Corps.
Discharged on June 27, 1865.
Died February 23, 1908.
Saw action at Chancellorsville.

Left: Obv. Whitehill. *Right:* Rev. Whitehill.

Private Reuben M. Post
Company K, 149th Pennsylvania Volunteer Infantry (Bucktails)

Mustered in on August 26, 1862.
Deserted February 14, 1863.
Also served in the 46th Pennsylvania Volunteer Infantry from September 13, 1861, until discharged on May 18, 1862.
Saw action at Kernstown, Virginia.

Private Jonathon Wood
Company B, 150th Pennsylvania Volunteer Infantry

Mustered in August 19 (23?), 1862, at Philadelphia, Pennsylvania, for 3 years as a private.
In Company B of 143rd Pennsylvania Volunteers, moved to 150th Pennsylvania Volunteers.

Left: Obv. Post. *Right:* Rev. Post.

Age 36, 5' 7½" tall, blue eyes, brown hair, light complexion, listed as a fuller.
Born in Huddersfield, England.
No record for August/September 1862.
Listed as present from October 1862 to May 1864.
Shown as prisoner of war July 1, 1863, Gettysburg.
Returned to duty from parole camp, West Chester, Pennsylvania, on August 8, 1863, having been illegally paroled at Gettysburg.
Listed as missing in action May 5, 1864, in the Battle of the Wilderness.
Dropped from the rolls September/October 1864.
Inventory of effects dated March 1, 1865, shows "No Effects."
Presumed killed in action.
Saw action at Gettysburg, Wilderness.

Left: Obv. Wood. Private Wood's identification disc was excavated in the area of the Wilderness battlefield where he was reported as missing in action. *Right:* Rev. Wood.

Private Mortimer H. Doud

COMPANY C, 151ST PENNSYLVANIA VOLUNTEER INFANTRY

Mustered in October 20, 1862, at Montrose, Pennsylvania, for nine months at age 22.
Died of disease at Union Mills, Virginia, on December 29, 1862.

Left: Obv. Doud. *Right:* Rev. Doud.

Sergeant Walter McCabe

Company B, 155th Pennsylvania Volunteer Infantry (Zouave)

Mustered in on August 23, 1862, at Pittsburg, Pennsylvania, for three years at age 21.
Promoted to corporal on December 26, 1862.
Promoted to sergeant on September 25, 1864.
Present for duty from muster in until May 1863.
In hospital until June 1863.
Present for duty until mustered out on June 2, 1865.
Saw action at Fredericksburg, Gettysburg, the Overland Campaign, Weldon Railroad, Peebles Farm, Hatcher's Run, and Five Forks, Virginia.

Left: Obv. McCabe. *Right:* Rev. McCabe.

Private Frederick W. Berg

Battery D, Pennsylvania Independent Artillery

Mustered into 14th Pennsylvania Volunteer Infantry for three months on April 27, 1861, from Reading, Pennsylvania.
Mustered out on August 27, 1861.
Mustered September 24, 1861, was attached artillery with 104th Pennsylvania Volunteer Infantry.

At muster in he was 27 years old, 5' 6" tall, had blue eyes, blonde hair, and had been employed as a plasterer.

Transferred to independent command, then Ninth Corps.

Reduced to private on February 3, 1863.

Absent without leave from March 29, 1863, until mid–April 1863.

Re-enlisted as a Veteran Volunteer on January 29, 1864.

Mustered out on June 13, 1865.

Saw action at Kelly's Ford, Virginia, 2nd Manassas, Chantilly, South Mountain, Antietam, Vicksburg, Jackson, Mississippi, Overland Campaign, and Petersburg.

Left: Obv. Berg. *Right:* Rev. Berg.

2nd Lieutenant Matthias Bitner

Company B, 2nd Pennsylvania Heavy Artillery

Enlisted Green Castle, Pennsylvania, January 26, 1863, as a private at age 19.

He was 5' 5" tall, had blue eyes and light hair, and was employed as an ambrotypist.

Present for duty from January 1863 until January 29, 1866.

Promoted to corporal on March 12, 1864, to sergeant on December 10, 1864, and to 2nd lieutenant on April 26, 1865.

Left: Obv. Bitner. Bitner's identification disc is also displayed in Lord's *Civil War Collector's Encyclopedia*, Vol. V, page 87. *Right:* Rev. Bitner.

Mustered out January 29, 1866.
Saw action at Petersburg and Chaffin's Farm.

Corporal Harrison B. Ward

COMPANY M, 2ND PENNSYLVANIA HEAVY ARTILLERY

Mustered in on August 6, 1862, at Pittston, Pennsylvania, for three years at age 20.
He was 5' 7" tall, had black eyes and a dark complexion, and was employed as a wheelwright.
Present for duty from muster in until mustered out on June 20, 1865.
Appointed corporal on November 11, 1864.
Saw action at Cold Harbor, Petersburg, the Mine Explosion, Weldon Railroad, and at Burmuda Hundred.

Left: Obv. Ward. *Right:* Rev. Ward.

Private Isaac A. Price

COMPANY M, 2ND PENNSYLVANIA VOLUNTEER CAVALRY

Mustered in October 1, 1861, at Pine Creek, Pennsylvania, for three years, at age 34.
He was 5' 10" tall with a fair complexion.
Enlisted in the Veteran Volunteers on December 17, 1863.

Left: Obv. Price. *Right:* Rev. Price.

Present for duty from muster in until declared missing in action on July 12, 1864, at Petersburg.
Captured at Petersburg and sent to Andersonville prison.
Paroled December 6, 1864.
Mustered out on July 13, 1865.
Saw action at 2nd Manassas, Occoquan, Virginia, Gettysburg, Parker's Store, the Wilderness, Todd's Tavern, Trevilian Station, and Petersburg.

Private James M. Shoop

COMPANY M, 3RD PENNSYLVANIA VOLUNTEER CAVALRY

Mustered in December 20, 1861, at Newville, Pennsylvania, for 3 years, as a private.
5' 6" tall, blue eyes, light hair, listed as a laborer.
Present from January 1862 to August 18, 1862.
Detached to guard train at Fort Monroe.
Returned September 1862.
Taken prisoner, Hartford(?) Church, November 28, 1862.
Returned February 21, 1863.
Absent March 1863 with dismounted men.
Returned, Dumfries April 10, 1863.
Present until October 17, 1863, dismounted, detached service.
Present November 1863 to June 1864.
Reenlisted as Veteran Volunteer January 1, 1864.
July 27, 1864, on detached service as orderly at City Point.
Assigned to General Patrick until April.
Transferred to 5th Pennsylvania Cav (?).
General court-martial April (?), 1864.
Assigned to hard labor for 9 months, September 15, 1865.
Mustered out December 15, 1865.
Dishonorable discharge December 15, 1865.

The tintype is believed to be that of trooper Price based upon representations of the major auction house from which it was purchased along with his identification disc, but the image bears no separate identification.

Left: Obv. Shoop. **Right:** Rev. Shoop.

Saw action at Yorktown, Williamsburg, Seven Days, Antietam, Union, Upperville, Manassas Gap, Hartwood Church, and Gettysburg.

Private Edward Hamson

Company D, 19th Pennsylvania Volunteer Cavalry

Mustered into the 20th Pennsylvania Emergency Volunteer Infantry on June 17, 1863.
Mustered out on August 10, 1863.
Mustered into the 19th Pennsylvania Cavalry on August 7, 1863, in Philadelphia.
Mustered out on June 17, 1865.
Saw action at Okalona, Mississippi, Cypress Swamp, Tennessee, Brice's Cross Road, Nashville, and Franklin.

Left: Obv. Hamson. *Right:* Rev. Hamson.

Corporal Joseph S. Wilson

Company L, 19th Pennsylvania Volunteer Cavalry

Mustered in on August 21, 1863, at Philadelphia, Pennsylvania, for three years.
Served fourteen months in detached service.
Mustered out May 14, 1866.
Saw action at Okalona, Mississippi, Brice's Cross Roads, Nashville, and Franklin, Tennessee.

Left: Obv. Wilson. *Right:* Rev. Wilson.

Rhode Island

Private Stephen A. Grover
COMPANY B, 11TH RHODE ISLAND INFANTRY

Mustered in on October 1, 1862, for nine months.
Present for duty until mustered out on July 13, 1863.
Mustered into the 3rd Rhode Island Volunteer Cavalry on October 13, 1863.
Captured on January 24, 1865, at Bayou Goula, Louisiana.
Mustered out on August 1, 1865.
Saw action at Suffolk, Virginia.

Left: Obv. Grover. *Right:* Rev. Grover.

Private Benjamin Cooper
COMPANY B, 1ST RHODE ISLAND LIGHT ARTILLERY

Enlisted on August 31, 1864, at Providence, Rhode Island, for one year, at age 28, as a substitute.
He had blue eyes and brown hair, was 5' 4" tall, and employed as a tailor.
Present for duty from September 1864 until mustered out on June 12, 1865.
Saw action at Hatcher's Run, Boydton Road and Fort Stedman.

Left: Obv. Cooper. *Right:* Rev. Cooper.

Vermont

Corporal Ebenezer Hopkins

Company E, 2nd Vermont Volunteer Infantry

Mustered in on June 20, 1861, at Turnbridge, Vermont, for three years.
Re-enlisted as a Veteran Volunteer on December 21, 1863.
Promoted to corporal on March 28, 1865.
Sick in hospital during April and May 1863.
Mustered out on July 21, 1865.
Saw action at 1st Bull Run, Peninsular Campaign, Antietam, Fredericksburg, the Overland Campaign, Opequon, Fisher's Hill, Cedar Creek, Petersburg, and Sailor's Creek.

Left: Obv. Hopkins. *Right:* Rev. Hopkins.

Private David Rollins

Company H, 3rd Vermont Volunteer Infantry

Enlisted on June 1, 1861, at Montpelier, Vermont, for 3 years, as a private, at age 22, 5' 6¾" tall, blue eyes, light hair.
Present September 1861.

Left: Obv. Rollins. *Right:* Rev. Rollins.

Present May 1862 until August 1862.
Absent in hospital September 1862.
Discharged for disability on November 13, 1862.
Enlisted February 21, 1865, in 7th Vermont Infantry.
Mustered out in July 1865.
Died May 27, 1904.
Saw action at Lewinsville, Lee's Mill, Williamsburg, Golding's Farm, Savage's Station, and White Oak Swamp.

Private James P. Elmer

COMPANY H, 5TH VERMONT VOLUNTEER INFANTRY

Mustered in September 16, 1861, at St. Albans, Vermont, for 3 years, as a private.
Age 22, 5' 8¾" tall, blue eyes, dark hair, dark complexion.
Occupation given as farmer.
Sick at muster in.
Absent November/December 1861 in Alexandria hospital.
Present January/February 1862 to March/April 1862.
Wounded in action on June 29, 1862, at Savage's Station.
Taken prisoner June 30 at James River.
In Richmond on July 13, 1862.
Paroled at City Point, Maryland, on August 3, 1862.
Wounds of right hip and abdomen (groin?) two or three wounds? (Not clear from records.)
Absent September/October in Hammond U.S.A. General Hospital at Point Lookout, Maryland.
September 15, 1862, detailed as nurse.
November/December 1862 detailed as wardmaster.
Absent till July/August 1864 when transferred to Invalid Corps, July 6, 1864.
Detached and sent to Auburn, New York, August 16, 1864.
Mustered out September 16, 1864.

A wartime image of Private Elmer (courtesy U.S. Army Military History Institute).

Left: Obv. Elmer. *Right:* Rev. Elmer. Private Elmor left his home town of "CANTON. N.Y." to enlist in a Vermont regiment.

Died February 15, 1913.
Saw action at Lewinsville, Lee's Mill, Williamsburg, Golding's Farm, Savage's Station.

Private Archibald Marston

COMPANY B, 6TH VERMONT VOLUNTEER INFANTRY

Enlisted September 19, 1861, or September 29, 1861, both dates are given in his service records.
Mustered in on October 15, 1861, at Bradford, Vermont, for 3 years, as a private.
Age 18, 5' 5½" tall, dark complexion, dark eyes and dark hair.
Born in Bradford, Vermont, either a paper maker or a farmer on different forms.
Status "not stated" November/December 1861 to February 28, 1862.
Present March/April to January/February 1863.
In January/February noted that "lost gun & equipments."
March/April 1863 absent sick in general hospital.
Present May/June 1863 to December 15, 1863.
December 15, 1863, enlisted as Veteran Volunteer. Due $15. Paid bounty of $60.
January 1864 to June 1864 present.
July/August 1864 present, noted as "sharp shooter Div HQ."
September/October 1864 to March/April 1865 present.
October 16, 1864, transferred to Company H.
Also listed as "A. Martin, A Marstin, A Mastin"
Mustered out June 26, 1865, at defenses of Washington, D.C.
Saw action at Yorktown, Williamsburg, Golding's Farm, Savage Station's, White Oak Swamp, Crampton's Gap, Antietam, Fredericksburg, Mayre's Heights, Gettysburg, Funkstown, Rappahannock station, Wilderness, Spotsylvania, Cold Harbor, Petersburg, Charlestown, Opaquan, Fisher's Hill, Cedar Creek.

Left: Obv. Marston. Not only was Private Marston's identification disc heavily worn, but it was apparently damaged in a fire at some point in its history. *Right:* Rev. Marston.

Wisconsin

Private John Giesbers

COMPANY E, 6TH WISCONSIN VOLUNTEER INFANTRY

Mustered in on July 16, 1861, for 3 years, as a private, age 28, at Camp Randle near Madison.
Present July 16 to August 31, 1861.

Not stated September/October 1861.
Present November/December 1861 to July/August 1862.
Listed as wounded on September 14, 16 and 17, 1862.
Also listed as deserted September 14, 1862.
Also listed as missing in action September 14, 1862.
War Department July 26, 1889: The charge of desertion of September 14, 1862 is removed and the record amended to read "missing in action at South Mountain, Md., September 14, 1862."
Saw action at Cedar Mountain, Gainesville (Brawner's Farm), and Bull Run, South Mountain.

Left: Obv. Giesbers. *Right:* Rev. Giesbers.

United States

Corporal Frederick Frese (Freese or Freeze)

COMPANY F, 1ST U. S. VOLUNTEERS

Mustered into the 20th Louisiana Infantry, C.S.A., on January 4, 1862.
Dropped from Confederate rolls on July 11, 1862 (probably captured and sent to Point Lookout POW Camp).
Enlisted into Federal service on April 27, 1864, at Norfolk, Virginia, and mustered in four days later.

Left: Obv. Frese. *Right:* Rev. Frese.

He was born in the free city of Bremen, Germany, and before enlisting was employed as a sailor.

At enlistment he was 5' 7" tall, and was described as having dark or gray eyes and light or dark hair.

In May and June 1865 he was absent, sick, and was given a disability discharge at Fort Snelling, Minnesota, on July 6, 1865.

Saw action in the Western Territories, Fort Wadsworth, Dakota Territory, and Fort Snelling, Minnesota.

Authentication

Collectors and dealers can expect to be occasionally offered misrepresented, fake, reproduction or fantasy Civil War identification discs. Because of this problem, authentication is of paramount concern to both. Even after years of observing hundreds of identification discs, authentication is never easy or certain, and not infrequently experienced disc collectors will have different opinions on the authenticity of minted identification discs such as those covered here.

So far, it appears that the large majority of miscreants who intend to deceive have limited their efforts to totally fake discs which can be marketed with a minimal investment and a high profit margin. Most of those are made either by using First World War or recent letter punch kits on old, heavily worn American or foreign copper coins which can be bought for a few dollars out of coin dealers cull boxes, or discs and tags made at home with sheet metal and shears. For those crooks who cannot even afford a punch kit or engraving tools, any sharp piece of metal seems to suffice to scratch on a name in hopes of cheating an unwary collector.

Not quite as prevalent as fakes deliberately intended to look authentic or as if made by a soldier in the field, are vintage pieces that were made for identification purposes after the Civil War but which are erroneously or knowingly mislabeled as war-time identification discs. It is unfortunately quite common to find misidentified luggage and other types of tags for sale in online auctions, or even at Civil War relic shows.

Not as common, but a concern nevertheless, are identification discs that were made by, or for, re-enactors. Although not intended to deceive, they occasionally make their way onto the market. Fortunately, most are either made from obviously reproduced discs, have tell-tale indicators, or bear the names of men who didn't fight in the Civil War.

In order to try to avoid bogus discs, most dealers and serious disc collectors have, in addition to the sixth sense that comes from experience, developed a mental check list of criteria and characteristics which are applied to every disc under consideration. To help new collectors, and as a reminder to old hands, the following suggested check list is offered.

Commercial Certification. Since about 1987, coin collectors have been assisted by the availability of coins which have been commercially authenticated and graded by experts whose opinions are backed by a certification and a guarantee. Once reviewed and approved, often by a team of graders, the coins are sealed in plastic containers with holographic labels to prevent fraud or tampering. Coins that have been certified and encapsulated are often referred to as "slabs." At this time there are approximately a half dozen companies that authenticate, grade and certify coins. One of them has expanded its service to include minted identification discs.

The Numismatic Guaranty Corporation of America, commonly referred to as NGC, began grading Civil War patriotic tokens and store cards in late 2001, and identification discs three years later. Along with PCGS and ANACS, NGC has become one of the more highly respected coin grading services.[1]

Certified and encapsulated identification discs are such a new development that the authors have had very limited opportunity to observe and evaluate them and therefore can only offer that no questionable discs have yet been observed in an NGC "slab." It can nevertheless be said, especially to novice collectors, that purchasing a certified and encapsulated identification disc offers an added degree of confidence and safety, in part because NGC claims that its authenticators have no financial interest in the transaction. As of this date however, the limited number of certified discs offered for sale has not provided a reliable indication on whether the concept will survive, and for the same reason the market has not yet determined whether already certified identifications discs will command any premium.

Provenance. Although extremely uncommon, the availability of authentic provenance such as reference to a described disc in a Civil War era letter, or depiction of same in an authentic image, should provide a very significant assurance of authenticity.

Soldier Records. If possible, it is very important to determine whether the name, regiment, company and other identifying information appearing on a disc matches that of an actual, verifiable Civil War soldier. If it does, the re-enactor problem is probably eliminated since presumably most re-enactors (or enthusiasts who purchase discs from modern sutlers at Civil War re-enactments) are more likely to have their own name inscribed on their disc rather than that of an actual veteran. Although not dispositive, the lack of a name match is a very strong signal for caution.[2]

It is recommended that any purchase of an identification disc be deferred until the soldier's information on the disc has been checked against the records, unless the seller provides an adequate return policy in case a discrepancy is discovered.

There are a number of places where the military records of a soldier can be reviewed. With sufficient time and the payment of a fee, the National Archives in Washington, D.C., will provide photocopies of a soldier's military and pension records.[3] Most states also have published lists of native sons who fought in the Civil War, usually by regiment. There is also a published registry for all Union soldiers that appears in a multi-volume set in some of the larger, or specialized, libraries.

The regiments for which a history was prepared often include a complete roster. Many such unit histories with rosters can be found on display at most shows and are potentially available for a quick check before a purchase is made. It is strongly recommended, though, that the owner's permission be requested and granted before a text is used solely as a reference.

Finally, there is soldier's information available through several online services. The United States government offers what is purported to be a complete list of all Civil War soldiers and sailors which can be accessed for free.[4] There is also at least one excellent commercial online data base available, which both authors frequently utilize to obtain soldier verification as well as information on Civil War regiments.[5]

Other information sometimes included on a disc about a particular soldier can be used as a tool for authentication. Chief among possible clues are; pertinent dates, the stated rank, the battles displayed and occasionally even the regiment.

An example of dates helping with authentication is found on the disc first owned by Private Edward Northrop, of the 31st New York Infantry.[6] His disc included the place and date "West Point May 7, 1862." That was the location and date of his regiment's first engagement with the Confederates.[7] Accurate information such as this is encouraging, while incorrect information should have the opposite effect.

The same method of analysis is useful if the soldier had his rank included on his disc. When Henry Donath of the 19th Massachusetts Infantry made his purchase, he had the sutler stamp on the rank of "Sgt. Major," along with the abbreviation "Fred'ksburg." Since Henry was promoted to Sergeant Major on October 14, 1862, and to Lieutenant on December 14, 1862, to replace another officer killed on December 13, 1862, at the battle of Fredericksburg, it seems extremely likely that he bought his disc sometime shortly after Fredericksburg but before his commission was approved and signed. If, for example, the information on this disc would have been conflicting, such as the rank not being correct or not making a logical fit, caution would have been in order.

As has been more thoroughly discussed earlier, some discs were minted specifically to include the places where a soldier saw action. It is strongly suspected that those style discs came with letter punch kits that had battles on punches as whole words, but in a much smaller font than that used for the soldier's name, company and regiment. Characteristically many of the longer battle names were arranged in an arc, either an inside or outside curve, to minimize the area consumed on the reverse.[8]

Other battles, even on the same disc, appear to have been punched one letter at a time. There is enough variation in the letter styles and their alignment to lead to the conclusion that some sutlers were either too frugal to buy separate battle punch kits if not originally included, or additional punches for battles that occurred after the punch kit was acquired. Nevertheless, the letters used to spell the names of the battles should be consistent with Civil War era letter styling in general (see the section on Lettering for a more extensive treatment of letter style) and hopefully with other lettering on the disc in question.

A good example of the mix of punches can be found on the identification discs of Asst. Surgeon Welsh of the 25th New York Infantry and Private Hull of the 37th New York. Each man had his sutler add several of the same battles in which their two regiments fought. The words "Before Richmond" are curved beneath the words "7 Days" on both. However, the stamps employed on each disc were slightly different in both size and shape. Further, the battle of Chantilly was spelled two different ways and small variations in the letter styles may also be discerned.

The majority of identification discs do not include battles, but among those that do the battles referenced were almost always fought in the East:

Presumed gang letter punches;
"Boydon Road [sic], Bristow Station, Bull Run (straight and curved), Cedar Mountain, Chancellorsville (straight), Chancellorsville (curved), Chantilla [sic], Chantilly, Coal Harbor [sic], Crampton Pass, Drainsville (curved), Fair Oaks, Fredericksburg (curved and straight), Fredericksb'g (curved), Fdksburgh, Gettysburg (curved and straight), Glendale, Goldsboro, Hatcher's Run, Kelley's Ford, Kingston, Locust Grove, Malvern Hill, Malv. Hill, Manassas Gap, Mine Run, New Berne, North Anna, Pensuliar Campaign [sic], Petersburg, Rappahannok Station [sic] (curved), Roanoke, Salem Church, South Mountain, S. Mountain (curved), Spottsylvania C.H. [sic], White Hall, Williamsburg, Wms'burgh, Winchester, Yorktown, 1st Bull Run, 2nd Bull Run, 7 Days Battles, 7 Days at Richmond, 7 Days Before Richmond, 7 Days Fighting Before Richmond;"[9]

Presumed individual letter punches;
"Antietam, Antetm [sic], Bottoms Bridge, Boyds Cross Road, Chancelorsville [sic] (curved with small letters), Chancelorsville [sic] (two lines), Charles City, Chickahomeny [sic], Cumberland Insland [sic], Fernandeno Fa., Fort Macon, Fort Pulaski, F'burg, Fredkburgh [sic], Fred'ksburg, Fredsburg December 13 1862, Gains Mills, Gainesville, Gettysburg, Gettysburgh [sic] Golden Farm (curved), Hampton Rodes [sic], Hanover Ct. House, Hanover C.H., Hattress [sic], Isle of Wight, Jackson, James Island, Malvern Hill, New Berne, Port Royal, Roanoke Island, Siege Yorktown, Sharpsburg, South Branch, S. Mills, So. Mountain, South Mtn, S. Moun, Suffolk, Tybe Island, Valso Inlet, West Point May 7, 1862, 7 Days Battle, 7 Days Bat's, 7 dy's Btl's on Pens'r, 7 dys. Bt .on Pn."

Not surprisingly, no discs have been observed that included Five Forks, Sailor's Creek or Appomattox, the last battles fought in the East.

On rare occasions a collector may encounter an identification disc which includes battles in which the soldier did not participate or which occurred even before he enlisted. Since all his comrades would have known when he joined, it is possible that either the sutler used some of his spare time to add the regiment's battles to a number of blanks for later convenience of sale, or the soldier thought it proper to list the battles fought by his regiment whether or not he was there. Despite possible explanations, such anomalies should raise significant concern unless authenticity is assured by more definitive indicators. If such an incongruity is coupled with some other problem, such as odd letter style, questionable surfaces, or an unusual word configuration, the disc should probably be avoided.

Several conclusions about the date of disc purchase, and therefore authenticity, may be gleaned from the specific battles included. In some cases a disc will include the last battle fought by a soldier before he was killed. Unless there is a variation in the letters indicating that battles were added serially, it can also be safely presumed that a disc was purchased at some time between the last battle listed and the next to be fought. If there is a gross error such as the disc including a battle that occurred after the soldier left the service or died — beware.

It can also be assumed that some sutlers, probably for a fee, were willing to stamp on additional battles after the fact. A distinct change in the lettering can disclose when the disc was obtained, or reveal the possibility of an alteration or outright fake if the stamped information is historically incorrect.

Several examples have been observed where it is likely the soldier had information added to his disc as events unfolded. Sergeant George Steele of the 30th New York Infantry Regiment seems to have had his rank added after his promotion to sergeant. Normally, the rank, when included, appears directly before the name along the curve. Probably because there was no room left from the first stamping, the rank was squeezed in elsewhere.

Another example is found on the disc of Private Henry Lown, who served in two regiments in the space of a year. When he bought the disc, the sutler stamped on his first unit, the 4th New York Heavy Artillery along with original enlistment date. A year after he was discharged from the 4th, he enlisted in the 16th New York Heavy Artillery and had the sutler add his new enlistment date and regiment.

Identification Disc Style. This work does not include home-made and/or engraved identification discs because the authenticity of same cannot be verified with any confidence. Any such item could have been made at home in 1862, or in 2002, with basically the same materials and technique.

Instead, the authors have attempted to provide a comprehensive catalog of all identification disc styles minted for Union soldiers. It is certainly possible, and probably likely, that disc sub-styles have either been missed or that others will be discovered. Caution is nevertheless recommended for any disc that deviates in any significant way from the examples shown, or which does not appear in Chapter 2.[10] If it appears that a new style has been discovered, and it can be purchased at a price which will justify the additional risk, good luck. Remember though, that luggage tags, political medals, souvenir shop material, and even hotel tags are sometimes misrepresented as Civil War identification discs.

Known Reproductions and Fakes. There are a number of reproductions, fakes and misrepresentations which any identification disc collector is likely to encounter. It would be impractical to try to catalog them all because the number and variety continually grows, but the following should provide a fair sampling of discs to avoid.

This work deals only with Union identification discs because almost all alleged Confederate discs are home-made and therefore nearly impossible to authenticate. There is one alleged

Confederate identification disc which requires some mention, however, because of its prevalence in the marketplace. Generally referred to as "Pender" discs after the most frequent name used, these reproductions are always made from either authentic or cast copies of moderately to heavily worn United States quarter, half or dollar coins of the seated liberty design. Frequently the coins are dated between 1860 and 1864.[11] On the reverse (eagle side) will usually appear an engraved, as opposed to stamped, name such as Dorsey Pender or James B. Washington, and a "C.S.A." on or directly below the eagle.[12] Unless one is interested in collecting all the variations of these fakes, they should be avoided.

Left: Reverse side of "Pender" counterfeit identification disc.

The items next most commonly misrepresented as Civil War identification discs, especially in online auctions, but occasionally on dealers' tables as well, are German Silver identification or luggage tags manufactured from as early as 1868 until at least the 1880s.[13] These tags (only one variety was perfectly round) consist of small medals, most commonly in the shape of a shield or an oval bearing a patriotic theme, especially an eagle in the background holding a scroll from its beak, on which the owners name and home town was to be stamped. It can be safely assumed these were manufactured and distributed after the Civil War since they were included in a mail order catalog dated 1886 from the firm of Peck & Snyder.[14] Avoid these even if regimental information is included since they were either made for a veteran who was proud of his unit or were intended to deceive.

Post–Civil War patriotic luggage tag, circa 1868.

A particularly deceptive reproduction, because it is minted in brass rather than cast, is sure to be more frequently encountered by collectors as time passes. Several years ago the Memorial Brass I.D. Manufacturing Co. of Argyle, New York, began marketing a brass copy of an Eagle 5A disc. For one price a blank could be purchased, but for

Left: Obverse side of modern reproduction identification disc marketed by Memorial Brass Mfg. Co.
Right: Reverse side of Memorial Brass reproduction identification disc.

an additional sum the company was willing to stamp the requested soldier's information on the reverse. Although the existence of such a quality reproduction is unfortunate, since it has and will continue to be a source of confusion, in fairness it must be mentioned that the manufacturer has deliberately included an identifying mark and perhaps intentionally allowed other identifying characteristics to remain.

The most obvious identifier is that even if bright, the Memorial's disc has not been gilded with gold, a fact which should be readily apparent unless heavily buffed to simulate extreme wear or perhaps artificially toned to simulate age. Almost as obvious, the tiny word "copy" appears on the reverse near the rim opposite from the suspension hole. Unfortunately, this marker is susceptible to obliteration by the kind of scratch or ding that is commonly seen on authentic discs.[15] Be especially wary of Eagle (5A) discs with a heavy scratch or other damage at exactly that point. Fortunately, there are other identifiers that are not so easily hidden.

On the obverse, the most easily detected clue appears on the arrows clutched in the Eagle's left claw (right side of disc). The top two of the three arrows are lacking a portion of their shafts where the die was insufficiently cut. Also, the reproductions, unlike the originals, have denticles rather than beads around the obverse and reverse rims.

Sometimes there are also bumps on the obverse corresponding to the letters stamped on the reverse which may be the result of too much force or a too soft of an alloy. The same excessive force sometimes even leaves the square face of the punch around a letter, a characteristic seldom observed on the genuine article.[16]

On the Memorial reproductions the stamped lettering on the reverse has both a distinctive appearance and arrangement. The soldier's name appears below the suspension hole, like on a Shield (2B) (which will be discussed below in greater detail), without the 90 degree rotation as is the norm for authentic Eagle (5A) discs. The letters lack the fancy serifs common to authentic discs, resembling more closely the letter style used on First World War discs. Finally, many of the letters tend to have an odd shape, heavy and rounded at the top and thin at the bottom, as if the punches were on an angle toward the hole when stamped.

Another category of bogus Civil War discs are those made using the First World War punch kits referred to above. This technique is most often encountered on hand-made copper discs or squares, or on old, heavily worn large copper cents or foreign coins.[17] Look at the serifs. In First World War kits, the E's, F's, L's and T's lack the characteristic triangle serifs in favor of straight serifs. Further, the World War I letters are about a half millimeter larger than the letters used on authentic discs.

Left: First World War letter and number punch kit for dog tags. *Right:* Inked letters and numbers from the World War I punch kit shown.

Collectors are also likely to encounter a post–Civil War restrike of the Washington (3D) style identification disc. Thomas Elder, a dealer in medals and tokens, somehow acquired the Civil War era obverse die for the Washington (3D) "Security" style disc. Perceiving a market, he commissioned a die-cutter named C.H. Hanson to make a die for the reverse. In mid–1917 Elder began minting, and in one of his numerous catalogs offering for sale, what he called a "Soldier's Identification Medal." The discs were minted in sterling silver, silver plate, bronze, brass and aluminum.

Elder's plan to replicate the marketing of Civil War identification discs apparently collapsed because the men received dog tags with the rest of their "Government Issue." As a result, very few of his discs are available with authentic World War I engraving or stamping, but many blanks still circulate.

These re-strike discs are very easy to identify. On the original Washington (3D) the reverse was minted in blank, and although very rare, authentic Civil War discs with sutler stamped soldier information on the reverse exist. On the Elder disc, though, Hanson added the word "Name" at the top, a blank underline beneath, another underline beneath that starting with "Co.," and a final underline starting with "Regt." At least one disc with accurate information from a Civil War soldier on an Elder restrike has been spotted for sale in an online auction. Presumably the information was added at or after the First World War (it is certainly possible that the soldier whose information appeared on the disc was a Civil War veteran who bought an Elder restrike and had it stamped with his accurate information as a memento, but it was not worn during the Civil War). If one is collecting only Civil War discs, Elder re-strikes should be avoided.[18]

Reverse side of Joseph Elder restrike, circa 1916.

A search of the internet or newspapers dedicated to Civil War enthusiasts will usually reveal a number of advertisements from modern day sutlers who cater to re-enactors. These merchants can also be found at most re-enactments selling reproduction equipment and gear to both re-enactors and spectators. Many modern sutlers offer reproduction identification discs modeled in design and size after the originals. Fortunately, these reproduction discs are almost always obviously cast and occasionally made of an inappropriate metal. For examples, one can find reproduction Eagle (5A) and McClellan (1A) discs cast in pewter or some similar white metal, and a few McClellan (1I), (1J) and (1K) discs have been found that were made from aluminum.[19]

In a category all by itself is a cast reproduction of a Shield (2B) that one of the authors picked up at a re-enactment. It is included for some comic relief.

There is one other reproduction Eagle 5A marketed by modern sutlers that could be very deceptive, especially to the novice disc collector. It is cast in brass without any of the identifiers found on the Memorial Brass reproduction. Warning clues are that there is no gilding, the shield on the eagle's breast is smeary and lacks detail, and the border on the obverse consists of denticles rather than

Obverse side of cast pewter reproduction of Eagle (5A) identification disc.

Left: Obverse side of cast pewter reproduction McClellan (1A) identification disc. *Right:* Reverse side of cast pewter reproduction McClellan (1A) identification disc.

Left: Obverse side of cast aluminum reproduction McClellan (1K) identification disc. *Right:* Reverse side of cast aluminum reproduction McClellan (1K) identification disc.

Obverse side of cast reproduction Shield (2B) identification disc.

beads. Both sides also have very small casting pits over the entire surface, which could be mistaken for corrosion marks. If observed in blank with a magnifying glass, one can also observe the remnants of letters from the original disc used to make the cast. Because none have been seen with punched information, it is not known whether the original letters will still be visible under or around punched letters.

Surface Appearance. The texture of a purported identification disc can often provide quick insight into its authenticity. All manufactured discs were made by a minting machine of one type or other. As a result, the surface will usually appear relatively smooth (except for occasional strike-overs, post manufacture blemishes or ground action on excavated pieces) and the lettering and image will usually appear crisp and sharp.[20] On all but the most

Left: Obverse side of cast brass reproduction Eagle (5A) identification disc. *Right:* Reverse side of cast brass reproduction Eagle (5A) identification disc.

heavily worn, the sides of the minted letters should appear to be fairly perpendicular to the disc's surface under a 10× magnifying glass.

All cast discs should be considered fakes or reproductions. There are four key indicators that usually allow identification of a cast disc, and the cheaper the effort the more readily apparent are the signs of casting.

The first is the appearance of the fields.[21] If they appear pitted, as if once boiling and then cooled leaving bubble craters, it is almost certainly a cast, because that is approximately what happened.

The second give-away is the quality of the strike. If the edges of the device and minted letters appear slanted or rounded, like a sand castle that has dried and the sand has begun to run down the vertical slopes, it is probably cast as well. Because of the limitations of the process, it is difficult to get an exact reproduction from a mold. Likewise, if the image appears smeary, or details are indistinct, the disc may have been poured.[22]

Sometimes heavy wear or a poor strike can also make the device on a minted disc smeary or indistinct. To the extent that it can be determined, the edges of the device should still be relatively vertical on an authentic but worn disc, because both a bad strike and wear take the highest portion of the image off first, and only extremely worn discs will show deterioration on the edges of the devices and letters.

One should be increasingly cautious in direct proportion to the amount of wear or loss of detail on a disc. Although minted identification discs were exposed to considerable genuine wear, artificially wearing down a cast or otherwise phony disc is a technique that will be encountered.[23] Remember, the sutler was a merchant trying to sell a relatively expensive product. Presumably, poor or shallowly struck discs would have been rejected, first by the maker, then by the sutler, and finally by his customers.

On the other hand, the presence of some wear, dings, dents, nicks or scratches provide a basic indicator of authenticity. The discs were made from fairly soft metal and came into contact with harder objects when; worn on a blouse or under a shirt, thrown into a jewelry box, or carried in the pocket. Few identification discs have been observed that are virtually blemish free, except for blanks that were never used. Unfortunately, the existence of blemishes is not proof of authenticity, but their unusual absence should provide a warning.[24]

Next, check the rim. The casting process requires a hole through which to pour or force the liquid metal. This entry point is usually on the rim, and if not corrected, a small nipple will remain. Usually the nipple is filed off, leaving tell-tale file marks in its place. Similarly, whether sand or plastic molds are used, the process generally leaves a ridge or thin raised track around the rim where the two sides of the mold were pressed together. If either a ridge on the rim or filing marks are present, the disc has a big problem.

A final clue sometimes appears on the reverse. Occasionally a disc will be observed with what appear to be very faint letters in the field, most often on the periphery near the rim. Apparently, some reproducers were so sloppy or rushed that they took their casting from an authentic disc, but failed to completely fill the soldier's information with wax or grease. When the original was pressed into the casting medium, some of the lettering remained, leaving faint raised letters on the casting.

More sophisticated crooks have made new dies to produce minted counterfeit coins rather than be satisfied with mere cast copies. Although no identification discs made from counterfeit dies have been recognized, a collector should still be wary of the potential for deception. The most easily detected products of fake dies are made by a method known as impact transfer where a genuine coin is smashed into a blank die. This method has an obvious negative impact on the detail of the new die.

A much more deceptive method is called spark erosion. This procedure involves submerging the original and the blank die in an electrolytic solution. Then an electric current is passed through, etching a nearly exact copy of the original into the die. Fortunately the etching process leaves tiny pitting on the die which then has to be buffed out. The final product generally has very smooth fields with very rough or bumpy devices, a situation not found on authentic identification discs.[25]

All of the foregoing methods would make it extremely difficult to reproduce the stamped soldier information from an authentic completed identification disc. Therefore, when analyzing the surface appearance, the disc, rather than the stamped information should be the focus. Stamping will be discussed in the Lettering section.

Color and Wear. Minted Civil War identification discs were manufactured primarily from brass or white metal (usually pewter and more rarely silver).[26] Most of the brass discs were gilded with a thin coating of gold, while those made from white metal were usually covered with a silver wash.[27] As a result, their coloration and shading usually falls into fairly consistent patterns and can provide a good indicator as to authenticity.[28] Because discs tend to darken with age and wear, the color should be fairly consistent with the amount of wear and the number of blemishes.[29]

Because the gilding held up so poorly, it is very common to see discs with some gilding on the fields surrounding the letters and devices but absent in the center of the fields and on the high spots.[30] This is because the latter received most of the contact and wear, while the areas on the fields closest to the lettering and the low spots in the device received far more protection from rub.[31] It is extremely unlikely to see gilding on top of the highest areas of the letters or device but worn away in the fields.

Although possible on very heavily worn discs, it is very unusual to find an authentic disc on which absolutely no gilding remains. Such a disc should be given special scrutiny because no reproductions have been observed that were coated with either gold or silver. Again, there should be consistency between the amount of wear and the remaining gilding and contact marks. If a disc has no gilt remaining, it should be very heavily worn and scuffed as well.

The silver coating also provides a different kind of clue on discs made from white metal. On that type it is normal to find a small circle of missing silver wash around all of the letters added by the sutler. The best guess as to the cause of this phenomenon is that the wash was so thin or brittle that the impact caused small portions of wash to flake off or to be damaged in such a way as to be more susceptible to wear. It is very common to find moderately or lightly worn Washington (3A) discs with small, dull gray halos around the punched letters. The appearance of same should be expected and taken as an indicator of authenticity.

Authentic discs with moderate to heavy wear, that have not been buffed or dipped, should have fairly consistent color. Brass discs should be approximately the color of chestnut, golden oak wood, or dark mustard. White metal discs should be some shade of gray, from light charcoal to battleship gray, with some appearing almost black in whole or in spots, especially the fields.

The foregoing does not apply very well to identification discs that have been excavated. The ground action caused by different soil compositions, moisture levels, and acidity will affect discs in different ways. In general, white metal seems to corrode more quickly and heavily than brass. Moderate to heavy corrosion should be taken as a good indicator of authenticity.

Ground action can often cause an accumulation of green corrosion on brass similar to verdigris on bronze statues. But patina can also be artificially mimicked. Dedicated collectors may wish to acquire a few modern reproductions or very heavily worn large cents from the first half of the 19th century and do their own comparisons to develop a sense of artificial coloring.[32]

Example of ground action on an excavated Washington (3A) identification disc that belonged to Pvt. Josiah Jones, 2nd N.H.V.I.

The issue of stains, spotting or otherwise inconsistent toning on a disc is a difficult matter to evaluate. On the one hand, across the spectrum of handling, from being lovingly cherished in a velvet lined jewelry box to lying at the bottom of a moldy chest in a wet basement for 140 years, it should be expected that at least a few discs would develop stains or odd toning. Because of the almost infinite number of variables, it is impossible to give any specific guidance in this regard other than oddly stained, colored or toned discs should be given extra scrutiny. When it is time for re-sale, prospective buyers will have the same concerns.[33]

Finally, one should not neglect the appearance of the hole in the disc.[34] The wear on the inside of the hole and around its edge should be consistent with the wear on the surface of the disc. It would be unlikely to find a heavily worn and nicked disc with a hole that has pristine, sharp edges, and vice versa. White metal discs seem to show more wear at the hole than those made of brass, probably because the latter were harder. Attempting to evaluate the amount of wear inflicted on a hole is complicated by the fact that some discs were suspended on heavily abrasive chains (including watch chains for years after the war), some on leather and some on ribbons or strings. It is not unusual to encounter discs with the hole shaped like a pear with the neck almost reaching the rim, and even with the rim slightly stretched out of round by the tugging of the cord. The best advice is to look for any peculiarity or inconsistency regarding the hole as compared to the rest of the disc.[35]

Lettering. The soldier's information added to a disc by the sutler offers a very important tool for authentication. Clues are provided by two factors. The sutler had to choose how to arrange the words on the reverse of the disc. The lay-out of the words, and the specific information included is so uniform among specific sub-styles, and across all styles and regiments in general as to suggest that the punch kits came with instructions or a diagram indicating how the information was to be arranged.[36] Most commonly, the soldier's name was placed on the inside of the rim bracketing the suspension hole. Next, in the upper center came the company, usually designated as for example, "Co. A." Just below the company designation, the regiment was added, often including the state such as "119th Reg. Pa. Vols." If the soldier's home town or county was included, it usually appeared at the bottom, along the rim, opposite the name. Any configuration differing radically from the foregoing should be viewed with caution.

Three exceptions provide excellent, but not dispositive, indicators of authenticity. Most Eagle (5A) identification discs have a peculiar word arrangement which is very different from all the rest. Instead of the name bracketing the hole, the entire alignment is usually rotated ninety degrees to the right so that the name circles the rim from a central orientation at 3 o'clock. The rest of the information is arranged as above with the same quarter turn rotation. This trait is so prevalent that any collector should be especially wary of any Eagle (5A) disc with a different arrangement.[37]

Similarly, many of the McClellan (1A) discs have the information rotated 180 degrees so that the name is centered around the rim at 6 o'clock from the suspension hole. The prevalence of that arrangement is not as uniform as with the Eagle (5A) discs, but its appearance should add confidence while its absence should inject a measure of caution.

Finally the Shield (2B) has the distinction of being the only disc observed so far where the soldiers name uniformly appears just below the suspension hole rather than as close to the perimeter as possible. Unfortunately, this configuration is similar to that used in the manufacture of one of the reproductions referenced above. Because of the uniqueness of the letter arrangement and the similarity to a reproduction, it would be natural, but incorrect, to avoid all discs of the Shield (2B) style.

Unfortunately, there are just enough variations caused by the stamping process that an unusual word arrangement should not constitute absolute grounds for rejection. Rather, oddly arranged words on the reverse of a disc should be viewed with additional skepticism, and perhaps justify a discount, until a review of the other authenticators either confirms or refutes the suspicion.

The style of the letters punched by the sutler can also provide an excellent diagnostic tool. Almost universally, the lettering style of punches used during the Civil War was far more ornate than the starker lettering style to which we have become accustomed. Almost all authentic Civil War identification discs have letters with serifs (i.e. small decorative appendages) at the bottom of letters, especially protruding from the ends of the arms of E's, T's, F's and L's. Very rarely will a collector encounter a letter style without serifs.[38]

Fortunately for collectors, Civil War era letter punch manufacturers added an additional embellishment to some of their letters that are virtually always an indicator of authenticity. Rather than be satisfied with a straight serif hanging from the arms of a T or E, or rising from the base of the L, those manufactures shaped the serif with an almost triangular shape connecting the end of the serif to the arm to which it was attached. If the triangular serifs are present on those letters, one's confidence should increase significantly. If absent, so should one's caution.

In certain instances, there are variations in the ornateness of the letters that are fairly uniform among particular styles of identification discs. For example, many of the Shield (2A)s have a distinctive and exaggerated foot on the leg of the R and a small projection from the stem of the J like a heel on a boot. Other styles such as the Eagle (5A) usually have a J without the projection.

Another fairly good indicator is the center arm of the E's. On the McClellan (1A), that serif is often a triangle. On Shield 2A's there is usually a vertical cross-piece at the end, while on Washington 3A's the E tends to have a small center arm with a triangle on the end. Three McClellan 1D's have been observed from two different regiments with all the A's having a missing section from the apex to the left side of the cross bar, indicating that the A punch came from the manufacturer with the defect, or it had a weakness that caused that section to quickly break off.

As is usually the case, incorrect indicators should not be assumed to be definitive proof of a fraud or a reproduction. It seems that there were minor variations in letter punches for the same style discs. Apparently sutlers also continued to use the same punch set after changing disc style, kits undoubtedly were lost from time to time and had to be replaced and perhaps it was necessary to occasionally replace individual letter punches that broke or were worn beyond continued use. To further complicate the situation, the strike of the punch can alter the appearance of the letter. If not exactly perpendicular to the disc, the serifs on the side of the lean will be exaggerated while those on the other side may be diminished or even disappear.

Another fairly reliable indicator appears on most Washington (3A) style discs. Apparently the punch kit that accompanied that sub-style contained a separate punch, or punches that either singly or when combined produced the phrase "War of 1861." That phrase was usually stamped somewhere near the center of the disc. Very few other styles have that particular stamped phrase. Also, the Washington (3A) kits seem to have included a small decorative flourish shaped like this "-<" that was then flipped to create a design like this "-< >-" which was used as decoration or a divider. Sometimes there was also a star or an x added in the center. No similar design has been observed on any other style disc.

Finally, if the disc in question is a McClellan (1A) or (1B), a Washington (3A), a Battles (4C), or a Lincoln (6A), be sure to check for the die-sinker's mark. On the McClellans and the Lincoln, the mark is incorporated under the shoulder of the bust. On the Washington, Merriam put his mark just below the bust but on the Battles he put it on the hanger tab. All observed examples of these have die-sinker marks. If one is not present, be warned.[39]

The die-sinker mark can also provide a clue as to whether the disc in question was minted or is a cast reproduction. On all but the most worn, especially the McClellan and Lincoln, the mark will be generally readable and fairly well preserved. It is very difficult to obtain fine detail on a cast copy however, so if the mark is present but unreadable, smudged or merely a tiny lump of metal, extra caution is warranted

Dealers. Along with a basic knowledge of how to authenticate an identification disc, a beginning collector should give some thought to the sources from which discs can be purchased. The issue of dealers is difficult because there are so many of them, from large operations down to collector/dealers displaying a few items at their first show. While addressing each dealer is impossible, and recommending any in particular is inadvisable (not because there aren't reliable ones, but because the authors don't know every dealer and prefer not to cast aspersions by omission), there are certain guidelines that may be helpful for the novice identification disc collector.

First, the size or quality of the inventory on display in the store or on the show table does not necessarily correlate to the dealer's experience or expertise when it comes to the topic of identification discs. The best approach, if possible, is first to go to a few shows and just observe, chat, and ask questions. Talk to the dealers who have discs for sale. In general, dealers are very friendly and love to talk about themselves, their material, and the Civil War in general, since most were enthusiasts and collectors before they became dealers. Ask about the particular disc for sale, identification discs in general, what the dealer collects or has collected in the past, and the number of discs he has handled. All the foregoing information is pertinent when trying to evaluate a dealer's level of expertise.

Chatting with the owner of a judged display or exhibit that includes identification discs can be especially informative. Since the exhibitor is not selling, it can be presumed that providing information is his goal. Further, it is likely that an exhibitor has significant expertise in the area of his display. Go ahead and grill him with all reasonable questions. It can get boring standing next to a display for hour after hour, and besides, the exhibitioner wouldn't be there if he didn't love the topic.

Should it become obvious that a dealer with an interesting identification disc knows little or nothing about the item, it is advisable to rely on one's own judgment if adequate to the task, but if not, to seek help or avoid the disc altogether. If any of the information the dealer provides is clearly false in any regard, don't buy unless confident that the disc is authentic.

If a friend or acquaintance with greater expertise in the subject is at the same show, don't be afraid to ask for a second opinion about a disc of interest. This also applies to any dealers with whom a relationship exists. Getting advice is a standard procedure at shows and generally won't, and definitely shouldn't, hurt any feelings. Explain what is intended and usually the dealer with the disc will agree to a full refund if the purchaser is not satisfied with the results of the second opinion. If the dealer won't agree, perhaps the purchase should be delayed until the consultant can do an on-site inspection.

Always inquire about the dealer's return policy, and always get a receipt or business card that includes the dealer's address and phone number.

Exercise caution if the dealer doesn't have any sort of information regarding the military record of the soldier whose name appears on the disc. Because a particular soldier's war record can have a profound effect on the desirability and value of a disc, it is likely that the dealer checked on it before he made his purchase. If the record is strong, he will flaunt it. If he claims not to know anything about the soldier, or that he hasn't had time to check into the record, the buyer may suffer if it is later determined that the soldier's name does not appear in the records as stamped, if the soldier was a deserter, or if he only served for a very short time.[40] A strong proven record is very desirable whereas a weak record can be a drag on the resale price. Conversely, if the soldier was with a famous unit, was killed in action, or was killed, wounded, captured or even fought at Gettysburg, be prepared to pay, and later sell, at a premium.

Another good reason to attend a few shows before buying is to get a basic understanding of the price range for identification discs. The disc market is thin, and disc prices can vacillate significantly over time, sometimes even as a result of a few serious collectors either buying heavily or standing on the sidelines. As a result, no specific price guide has been included with this work.

It can generally be assumed that prices at a show will seek their proper level due to market forces. If a disc is encountered that is significantly higher in price than comparable discs for sale at the show, ask the dealer for justification.

It is also worth remembering that the dealers get to see each other's merchandise before a show is opened to the public. A good piece priced too cheaply will most likely be snapped up before collectors get a chance at it. One should be careful about a disc that seems to be very inexpensive compared to similar items on the floor.[41]

Online Sales. Buying identification discs online, whether through auctions or from dealer web-sites, presents additional concerns. Practically anyone can sell anything in an online auction with almost any imaginable description. At any given time there are usually items for sale in online auctions which experts would agree are either misrepresented as Civil War relics or are even deliberate fakes.

Before buying from an online auction, a certain amount of reconnaissance of that medium is strongly suggested. In particular there are a number of web-sites that deal specifically with online auction fraud, and reviewing them, especially the one provided by the American Relic Hunters, can be an educational and eye opening experience.

After one has gained a general sense of what a jungle online auctions really are, spend a month or so just watching and learning. Because they are so rare, one will usually only find one or two authentic minted identification discs available during any particular week, and it is not unusual for several weeks to pass without any. Watch the bidding and the prices realized, and never bid unless you are confident the item is authentic. When it is time to sell, the market will punish mistakes, and even questionable discs, without mercy.

If it is decided that a disc appears authentic enough to justify further consideration, and in this regard don't be afraid to post an inquiry on the American Relic Hunters "fakes" site (there is incredible expertise available, and when there is an error, it is usually on the side of caution), the next question should be whether the method of sale or the seller raises cause for concern. Be wary of a disc offered by a seller who is holding a private auction.

The same advice holds true for sellers who hide their feedback or feedback rating. It is likely they are doing so for a reason. And in general, the lower the rating below 99.5 percent the more caution should be exercised.

If the seller's pictures are bad, either too small, or too blurry to read, request better pictures. If not forthcoming, be careful. It can be safely assumed that a seller with a good piece wants it seen in all its glory and detail. There are instances where a seller may lack the technology or the initiative to take good pictures (especially an amateur who isn't a relic dealer), but because identification discs are fairly high end relics, those sellers are rare. If the item can't be seen clearly, one should wonder if that is the effect the seller intended.[42]

Despite the foregoing, online auctions probably offers the best market to find authentic Civil War identification discs. It is also the most risky. To borrow a phrase, "be careful out there," and always remember there will be another disc next week.

In a similar vein, it is wise to remember that there are no officials who issue licenses or otherwise regulate online web-sites. Fortunately, very few problems have been observed to date involving online dealer fraud or misrepresentation regarding identification discs.[43] As further assurance, many dealers with online web-sites also sell at shows, and thus the same natural policing tends to occur.

In general, however, one should apply the same tests for online web dealers as for dealers at shows, paying particular attention to seriously underpriced discs and questionable representations. Given the necessity of buying without actually taking delivery of the disc at the time of payment, it would be prudent to obtain a street address from an online dealer who only lists a post office. It is also always advisable to confirm an online dealer's return policy before making a purchase if not provided on the site, make a copy of the page on which the disc appears so as to be able to make a comparison and as proof, and to reference the return policy in the payment letter.

* * *

In summary, it must be emphasized that none of the indicators listed above, except for the signs of casting and similarity to known reproductions, should be used solely as the basis for rejecting a particular disc. Even the United States Mint generated hundreds of variations on its coins during the same year until the late 19th century when minting technology improved. Unlike the mint, sutlers operated in much more trying and varying circumstances, with varying degrees of skill and concern, and with little impetus for consistency.

Rather, the indicators above should be used like a check list. The greater number of consistencies, the higher the likelihood that the identification disc in question is authentic. Conversely, the larger the number of question marks, the greater the need for caution until a point is reached when the disc ought to be rejected.

Authentication Quick Check

1. General
   - ✓ Is there any misidentified material on dealer's table?
   - ✓ Does the price seem too good to be true?
   - ✓ Do the records support the disc?
2. Appearance
   - ✓ Are there any signs of casting?

✓ Is there any resemblance to know reproductions?
Are portions of the arrow shafts missing on Eagle (5A)?
Are there any suspicious gouges at criticial locations?
If an Eagle (5A), are there beads around the rim?
- ✓ Are toning, wear and surface abrasions consistent?
- ✓ On brass disc, does any gold gilding remain?
- ✓ On non-brass disc, does any silver wash remain?
- ✓ On non-brass disc, are there halos around stamped letters?
- ✓ Does the die-sinker's name or initial appear appropriately?
- ✓ Is the metal incorrect for the disc style.

3. Lettering
- ✓ Is the information on the reverse appropriately rotated?

180 degrees for a McClellan (1A) disc.
90 degrees right for an Eagle (5A).
- ✓ Do the key letters E, F, L and T have correct serifs?
- ✓ Does the Washington (3A) have correct symbols?

A design that looks like -< >- or -<x>-.
A stamped "War of 1861."
- ✓ Is information stamped in the proper location?

Name around the edge and straddling the hole.
On a Shield (2B) just below suspension hole.
Are the company and regiment in the central area?
Is the city or county around the edge opposite the name?
- ✓ Is the lettering unusually sloppy?

⑤

SURVEY OF IDENTIFICATION DISCS BY STATE AND STYLE

In this chapter, the reader will find two tables that present information on 615 identification discs that the authors own, have see, are in collections owned by experts, or have been observed in various formats such as respected texts, traditional auction catalogs, online auctions or shows. These tables provide a reference for the reader regarding known Union regiments and the styles of discs the soldiers in those units owned.[1]

The tables are sorted in two ways. In the first table identification discs from all the sources listed above have been sorted by state, type of unit, regiment, company, and disc substyle. This arrangement allows the reader to quickly check to see, for example, if the 46th New York Infantry was known to have any identification discs; what particular substyle, if any, was favored, and even if one company chose a different substyle than other companies in that regiment. The second table presents the same information sorted by style and substyle of the various identification discs, and then repeats the information for state, regiment company and type. This allows the reader to judge the relative number of discs of any particular style or substyle that the authors have observed and which units are known to have used this type of disc.

The column labeled "source" provides a basic caveat as to the reliability of the information by indicating that those discs marked by the designation *** were observed only on an online auction. Although the authors have a fairly high degree of confidence, it must be remembered that their decisions on those discs were based solely on the photographs observed. All other discs came from the other sources, of which the authors have a much higher degree of confidence. The last column provides information on the rank or presentation of the soldier as shown on his disc

A. Arranged by State

State	Type	Unit	Company	Disc style	Source	Rank
California	Infantry	1st	B	1A	***	
Connecticut	Infantry	8th	B	1B		
Connecticut	Infantry	8th	K	5C		
Connecticut	Infantry	8th	F	1A		
Connecticut	Infantry	8th	G	3B		
Connecticut	Infantry	11th	D	5C		
Connecticut	Infantry	11th	D	5C		
Connecticut	Infantry	11th	G	2A		
Connecticut	Infantry	11th	none	5C		
Connecticut	Infantry	12th	K	3B		
Connecticut	Infantry	14th	H	3C		

State	Type	Unit	Company	Disc style	Source	Rank
Connecticut	Infantry	15th	G	5C		
Connecticut	Infantry	18th	E	3D		
Connecticut	Infantry	18th	E	3C		
Connecticut	Infantry	18th	B	3C		
Connecticut	Infantry	19th	E	4B		
Connecticut	Infantry	21st	C	3B		
Connecticut	Infantry	21st	F	3C		
Connecticut	Artillery	1st Heavy	D	1C		
Connecticut	Artillery	1st Heavy	D	1C		
D.C.	Infantry	2nd	C	10A		
D.C.	Infantry	2nd	E	3A		
Delaware	Infantry	1st	I	2A		
Delaware	Infantry	2nd	G	1D		
Delaware	Infantry	2nd	B	5A		
Delaware	Infantry	2nd	E	5A		
Delaware	Infantry	2nd	H	5A	***	
Delaware	Infantry	2nd	H	5A		
Delaware	Infantry	2nd	C	1B		
Illinois	Infantry	58th	?	5A		Sgt.
Illinois	Infantry	59th	?	6A silver		
Illinois	Cavalry	12th	A	6A	***	
Indiana	Infantry	9th	C	4C	***	
Indiana	Infantry	17th	B	4B		
Indiana	Infantry	20th	F	16A		
Indiana	Infantry	20th	C	3A	***	
Indiana	Infantry	20th	H	4C		
Indiana	Infantry	20th	H	4B		
Indiana	Infantry	20th	?	16A	***	
Indiana	Infantry	130th	I	9A	***	
Indiana	Infantry	147th	B	4B		
Indiana	Cavalry	3rd	B	3A		
Maine	Infantry	3rd	A	3A		
Maine	Infantry	3rd	E	1L?		
Maine	Infantry	3rd	E	1L?		
Maine	Infantry	7th	F	5A	***	
Maine	Infantry	7th	H	5A	***	
Maine	Infantry	17th	B	4B		
Maine	Infantry	19th	G	3D		
Maine	Infantry	27th	G	1C		
Maine	Artillery	1st HA	H	4B		
Maine	Artillery	1stHA	H	4B		
Maryland	Infantry	1st Md PHB	A	?	***	
Maryland	Infantry	1st Md PHB	I	2C		
Maryland	Infantry	2nd	F	1A	***	
Maryland	Infantry	7th	D	2A	***	
Maryland	Infantry	10th	A	2C		
Maryland	Infantry	10th	A	2C		
Maryland	Cavalry	1st	G	6A		
Massachusetts	Infantry	1st	H	3A		
Massachusetts	Infantry	1st	C	3C		
Massachusetts	Infantry	1st	K	4C	***	
Massachusetts	Infantry	2nd	G	5A	***	
Massachusetts	Infantry	3rd	C	3C		
Massachusetts	Infantry	3rd	C	3C		Captain
Massachusetts	Infantry	7th	C	1D		
Massachusetts	Infantry	7th	E	? No picture	***	
Massachusetts	Infantry	8th	E	6A	***	
Massachusetts	Infantry	9th	B	2A		Sgt.

State	Type	Unit	Company	Disc style	Source	Rank
Massachusetts	Infantry	9th	G	2C?		
Massachusetts	Infantry	10th	D	3A		
Massachusetts	Infantry	12th	B	5A	***	
Massachusetts	Infantry	16th	I	1A		
Massachusetts	Infantry	16th	K	4C		
Massachusetts	Infantry	19th	1A			Sgt. Maj.
Massachusetts	Infantry	19th	B	1A		
Massachusetts	Infantry	20th	?	4C		
Massachusetts	Infantry	21st	G	1A	***	
Massachusetts	Infantry	23rd	G	1D?		
Massachusetts	Infantry	28th	?	1H?	***	
Massachusetts	Infantry	28th	H	5A	***	
Massachusetts	Infantry	32nd	A	2A	***	
Massachusetts	Infantry	32nd	H	6A		
Massachusetts	Infantry	34th	A	4B	***	
Massachusetts	Infantry	35th	E	3A	***	
Massachusetts	Infantry	41st	E	6A	***	Lt.
Massachusetts	Infantry	41st	A	6A		bugler
Massachusetts	Infantry	42nd	I	3A	***	
Massachusetts	Infantry	44th	I	3A		
Massachusetts	Infantry	53rd	K	2B		
Massachusetts	Infantry	56th	H	3A	***	
Massachusetts	Infantry	58th	A	2B	***	
Massachusetts	Infantry	59th	I	3A	***	
Massachusetts	Cavalry	1st	K	3A		
Massachusetts	Cavalry	1st	K	3A		
Massachusetts	Cavalry	4th	G	3A	***	
Massachusetts	Artillery	9th Batty.	A	3B?		
Massachusetts	Artillery	4th Heavy	M	9A		
Michigan	Infantry	1st	E	2B		
Michigan	Infantry	1st	G	2?	***	
Michigan	Infantry	2nd	C	3A	***	Drummer (dug)
Michigan	Infantry	2nd	I	3A		
Michigan	Infantry	3rd	A	4C		
Michigan	Infantry	3rd	I	1A		
Michigan	Infantry	4th	K	1C	***	
Michigan	Infantry	6th	C	6A		
Michigan	Infantry	8th	?	5A		sutler
Michigan	Infantry	8th	B	1A		sutler
Michigan	Infantry	18th	F	4A		
Michigan	Infantry	26th	A	5B		
Michigan	Cavalry	6th	C	6A		
Michigan	Cavalry	6th	C	6A		
Michigan	Cavalry	6th	H	6A	***	
New Hampshire	Infantry	2nd	?	5A		
New Hampshire	Infantry	2nd	C	3A		
New Hampshire	Infantry	2nd	H	3A		
New Hampshire	Infantry	2nd	E	3A		
New Hampshire	Infantry	2nd	E	3A		
New Hampshire	Infantry	2nd	F	4C		
New Hampshire	Infantry	4th	G	6A	***	
New Hampshire	Infantry	5th	A	1B		
New Hampshire	Infantry	5th	B	5A	***	
New Hampshire	Infantry	5th	B	5A		
New Hampshire	Infantry	5th	D	5A		Corporal
New Hampshire	Infantry	5th	F	5A	***	
New Hampshire	Infantry	5th	I	5A	***	

State	Type	Unit	Company	Disc style	Source	Rank
New Hampshire	Infantry	5th	H	5B?	***	
New Hampshire	Infantry	5th	H	5B?	***	
New Hampshire	Infantry	6th	B	5A	***	
New Hampshire	Infantry	6th	I	5A		
New Hampshire	Infantry	6th	I	12A		
New Hampshire	Infantry	8th	E	1A	***	
New Hampshire	Infantry	9th	B	2D		
New Hampshire	Infantry	9th	B	5A	***	
New Hampshire	Infantry	9th	C	5A	***	
New Hampshire	Infantry	9th	C	5A	***	
New Hampshire	Infantry	9th	D	5A	***	
New Hampshire	Infantry	9th	D	5A		
New Hampshire	Infantry	9th	F	5A		
New Hampshire	Infantry	10th	A	5A		Sgt.
New Hampshire	Infantry	10th	B	5A		
New Hampshire	Infantry	10th	D	5A		
New Hampshire	Infantry	10th	H	5A		
New Hampshire	Infantry	10th	H	5A		
New Hampshire	Infantry	10th	H	5A		Sgt.
New Hampshire	Infantry	10th	H	5A		
New Hampshire	Infantry	10th	H	5A		Sgt.
New Hampshire	Infantry	10th	I	5A		
New Hampshire	Infantry	11th	C	5A		
New Hampshire	Infantry	11th	D	5A	***	
New Hampshire	Infantry	11th	E	5A		
New Hampshire	Infantry	11th	E	5A		cook
New Hampshire	Infantry	11th	B	5A		cook
New Hampshire	Infantry	11th	H	5A		
New Hampshire	Infantry	11th	F	5C		
New Hampshire	Infantry	11th	F	5C		
New Hampshire	Infantry	11th	I	5C		
New Hampshire	Infantry	11th	I	5A		
New Hampshire	Infantry	11th	I	5A		
New Hampshire	Infantry	12th	B	5A		
New Hampshire	Infantry	12th	D	5A		
New Hampshire	Infantry	12th	D	5A		
New Hampshire	Infantry	12th	F	5A		
New Hampshire	Infantry	12th	G	5A		
New Hampshire	Infantry	12th	G	5A		
New Hampshire	Infantry	12th	I	5A	***	
New Hampshire	Infantry	12th	I?	5A	***	
New Hampshire	Infantry	13th	B	5A		
New Hampshire	Infantry	13th	C	5A	***	Corporal
New Hampshire	Infantry	13th	D	5A		
New Hampshire	Infantry	13th	H	5A		
New Hampshire	Infantry	13th	K	5A		
New Hampshire	Infantry	13th	E	1C		
New Hampshire	Infantry	14th	A	5A		
New Hampshire	Infantry	14th	B	5A		
New Hampshire	Infantry	14th	C	5A?	***	Corporal
New Hampshire	Infantry	14th	C	5A		
New Hampshire	Infantry	14th	D	5A		
New Hampshire	Infantry	14th	F	5A		
New Hampshire	Infantry	14th	G	5A	***	
New Hampshire	Infantry	14th	H	5A		Corporal
New Hampshire	Infantry	14th	H	5A		
New Hampshire	Infantry	14th	H?	5A		
New Hampshire	Infantry	14th	I	5A?	***	

State	Type	Unit	Company	Disc style	Source	Rank
New Hampshire	Infantry	14th	K	5A		
New Hampshire	Infantry	14th	G	1A		
New Hampshire	Infantry	15th	D	1A?	***	
New Hampshire	Infantry	15th	A	5A	***	Sgt.
New Hampshire	Infantry	15th	B	5A	***	
New Hampshire	Infantry	15th	C	5A	***	
New Hampshire	Infantry	15th	C	5A		
New Hampshire	Infantry	15th	G	5A	***	
New Hampshire	Infantry	15th	I	5A	***	
New Hampshire	Infantry	15th	I	5A	***	
New Hampshire	Infantry	15th	I	5A	***	
New Hampshire	Infantry	15th	K	5A	***	
New Hampshire	Infantry	15th	K	5A		
New Hampshire	Infantry	15th	H	5A	***	
New Hampshire	Infantry	15th	D	5B	***	
New Hampshire	Infantry	15th	C	?	***	
New Hampshire	Infantry	15th	G	5A	***	
New Hampshire	Infantry	15th	I	5A		
New Hampshire	Infantry	15th	I	5A		
New Hampshire	Infantry	15th	D	5A		
New Hampshire	Cavalry	1st	L	5A		
New Hampshire	Artillery	1st	?	5A		
New Hampshire	Artillery	1st	?	5A		
New Hampshire	Artillery	1st Batt. Vet. Vol.	?	16A		
New Jersey	Infantry	1st	F	1C		
New Jersey	Infantry	1st	F	3D		
New Jersey	Infantry	5th	E	4C		
New Jersey	Infantry	6th	G	3C		
New Jersey	Infantry	6th	A	3A		
New Jersey	Infantry	6th	G	3A		
New Jersey	Infantry	7th	H	4C		
New Jersey	Infantry	7th	none	11A	***	
New Jersey	Infantry	10th	B	1A		
New Jersey	Infantry	11th	D	4B	***	
New Jersey	Infantry	14th	A	5A	***	
New Jersey	Infantry	21st	E	1C	***	
New Jersey	Infantry	33rd	D	3C		
New York	Infantry	1st	I	5A		
New York	Infantry	2nd	C	3A		
New York	Infantry	2nd	E	3A		Sgt.
New York	Infantry	2nd	G	6A		
New York	Infantry	2nd	G	16A		
New York	Infantry	2nd	I	3A		
New York	Infantry	2nd sm (82nd)	I	2A	***	
New York	Infantry	5th	G	4C		
New York	Infantry	6th	A	3A		
New York	Infantry	6th	?	10A		
New York	Infantry	7th	K	5A	***	
New York	Infantry	8th SS	?	10A		
New York	Infantry	9th	A	6A	***	
New York	Infantry	11th	K	3B?		
New York	Infantry	13th NG	D	1B		
New York	Infantry	13th	A	1B		Captain
New York	Infantry	14th	A	1B		
New York	Infantry	14th Brklyn.	D	2A	***	
New York	Infantry	14th Brklyn.	E	2A	***	

State	Type	Unit	Company	Disc style	Source	Rank
New York	Infantry	15th Eng.	L	4B		
New York	Infantry	17th	H	2C		
New York	Infantry	17th	H	2C		
New York	Infantry	20th	B	1D		
New York	Infantry	21st		3A		
New York	Infantry	22nd	C	1B	***	
New York	Infantry	22nd	D	1B		
New York	Infantry	22nd	E	1B		
New York	Infantry	25th	C	4B	***	
New York	Infantry	25th	D	4B	***	
New York	Infantry	25th	?	4C		
New York	Infantry	30th	G	2A	***	Asst. Surg.
New York	Infantry	30th	G	2A		
New York	Infantry	30th	E	2A		Sgt.
New York	Infantry	31st	A	1C		
New York	Infantry	31st	A	5A	***	
New York	Infantry	31st	A	5A		
New York	Infantry	33rd	B	5A		
New York	Infantry	33rd	B	5A		
New York	Infantry	33rd	G	5A	***	
New York	Infantry	34th	D	2A		
New York	Infantry	34th	K	2A		Sgt.
New York	Infantry	37th	I	4B		
New York	Infantry	40th	E	4C		
New York	Infantry	43rd	D	1C		
New York	Infantry	43rd	E	1A		
New York	Infantry	50th Eng.	C	13A		Drummer
New York	Infantry	50th Eng.	?	13A		
New York	Infantry	50th Eng.	B	2A		
New York	Infantry	51st	B	1C		
New York	Infantry	51st	C	1F		
New York	Infantry	52nd	G	3A		
New York	Infantry	56th	K	8A		
New York	Infantry	57th	F	5A		
New York	Infantry	57th	H	5A		Lt.
New York	Infantry	57th	H	5A		Sgt.
New York	Infantry	57th	C	1B		
New York	Infantry	59th	I	2C	***	
New York	Infantry	61st	A	5A	***	
New York	Infantry	61st	F	5A	***	
New York	Infantry	61st	F	5A		
New York	Infantry	63rd	A	1C		
New York	Infantry	63rd	A	5A		
New York	Infantry	63rd	F	5A		
New York	Infantry	64th	G	5A		
New York	Infantry	66th	H	1B		
New York	Infantry	69th	E	2A	***	
New York	Infantry	69th	B	5A	***	
New York	Infantry	69th	F	5A	***	
New York	Infantry	69th	I	5A	***	
New York	Infantry	70th	D	3A		
New York	Infantry	71st	C	3A		
New York	Infantry	72nd	D	4B		
New York	Infantry	72nd	E	3A	***	
New York	Infantry	74th	D	4B		Wagon Master
New York	Infantry	74th	I	3A		Sgt.
New York	Infantry	74th	I	3A		

State	Type	Unit	Company	Disc style	Source	Rank
New York	Infantry	76th	A	5A	***	
New York	Infantry	76th	E	5A		
New York	Infantry	76th	H	5A		
New York	Infantry	76th	I	5A		
New York	Infantry	77th	C	3A		
New York	Infantry	80th	?	8A		
New York	Infantry	82nd	I	2A	***	
New York	Infantry	87th	C	3A		
New York	Infantry	87th	I	3A		
New York	Infantry	88th	H	1B		
New York	Infantry	89th	K	5B?		
New York	Infantry	95th	B	2C		
New York	Infantry	96th	C	2B		Sgt.
New York	Infantry	97th	D	2C		
New York	Infantry	99th	C	5C	***	
New York	Infantry	99th	F	5C		
New York	Infantry	101st	A	3B	***	Drummer
New York	Infantry	101st	F	3C		
New York	Infantry	104th	G	6A	***	
New York	Infantry	107th	H	6A	***	
New York	Infantry	107th	I	1A/B?	***	
New York	Infantry	108th	K	1A	***	
New York	Infantry	108th	K	1A		
New York	Infantry	120th	C	4C		
New York	Infantry	121st	I	2A		
New York	Infantry	124th	H	6A	***	Sgt.
New York	Infantry	125th	H	16A	***	
New York	Infantry	140th	G	2A		
New York	Infantry	140th	H	2A	***	
New York	Infantry	141st	K	1H		
New York	Infantry	145th	D	3C		
New York	Infantry	145th	I	1C		
New York	Infantry	145th	I	1C		
New York	Infantry	147th	E	2C		
New York	Infantry	147th	I	10A		
New York	Infantry	153rd	A	2A?		
New York	Infantry	153rd	D	1C		
New York	Infantry	153rd	I	1C		To Mother
New York	Infantry	153rd	E	1C		
New York	Infantry	156th	G	2B	***	
New York	Infantry	157th	A	1C		
New York	Infantry	159th	H	2C?	***	Presented to
New York	Infantry	164th	C	2A		
New York	Infantry	165th	C	13A		
New York	Infantry	169th	F	2A	***	Sgt.
New York	Infantry	169th	H	3B		
New York	Infantry	169th	G	2A		Drummer
New York	Infantry	169th	K	2A		corporal
New York	Infantry	Excelsior	?	14A		
New York	Infantry	3rd Excelsior	D	4B		
New York	Infantry	5th Excelsior	D	4B		
New York	Infantry	NYSS	?	10A		
New York	Infantry	NYSS	?	10A		
New York	Cavalry	1st Mtd. Rifles	K	4B		
New York	Cavalry	1st	H	6A		
New York	Cavalry	1st Vet.	B	1D		
New York	Cavalry	1st Vet.	C	4B		
New York	Cavalry	1st Vet.	C	1C	***	

State	Type	Unit	Company	Disc style	Source	Rank
New York	Cavalry	1st Vet.	I	1C	***	presentation to lady
New York	Cavalry	1st Vet.	M	1C		
New York	Cavalry	2nd	A	6A		
New York	Cavalry	2nd	B	2A		
New York	Cavalry	2nd	E	1D		
New York	Cavalry	2nd	G	2A		
New York	Cavalry	2nd Vet.	C	1D		
New York	Cavalry	2nd FZ	K	3A		
New York	Cavalry	2nd Mtd. Rifles	L	6A	***	
New York	Cavalry	Harris Light (2nd)	F	3A		
New York	Cavalry	Harris Light (2nd)	L	3A		
New York	Cavalry	3rd	F	5D	***	
New York	Cavalry	3rd	L	5D	***	
New York	Cavalry	3rd	G	5C	***	
New York	Cavalry	4th	H	10A		
New York	Cavalry	9th	H	2A	***	
New York	Cavalry	10th	A	6A	***	
New York	Cavalry	10th	L	5D	***	
New York	Cavalry	15th	K	1C		
New York	Cavalry	18th	D	1C		
New York	Cavalry	18th	E	1C		
New York	Cavalry	18th	G	1C		
New York	Cavalry	18th	G	1C		
New York	Cavalry	20th	B	1D		
New York	Cavalry	21st	A	1A?		
New York	Cavalry	21st	C	1C	***	
New York	Cavalry	21st	H	1C		
New York	Cavalry	21st Vet.	A	1C		Corporal
New York	Cavalry	24th	H	1C		
New York	Artillery	1st	C	5A		
New York	Artillery	1st	G	1B		
New York	Artillery	1st	L	5A		
New York	Artillery	1st Light	D	3C	***	
New York	Artillery	2nd	A	1A		
New York	Artillery	2nd	B	1A		
New York	Artillery	2nd Heavy	M	1D		
New York	Artillery	3rd	E	4B?		
New York	Artillery	4th Heavy	H	16A	***	
New York	Artillery	6th	L	1F		
New York	Artillery	6th Ind. Batty	?	4B?		
New York	Artillery	13th Heavy	A	5B		
New York	Artillery	13th Heavy	A	5B		
New York	Artillery	13th Heavy	D	5D		
New York	Artillery	16th Heavy	A	16A		
New York	Artillery	16th Heavy	M	5C	***	
New York	Artillery	19th	D	1A		
New York	Artillery	Rock Batt	B	3C		
Ohio	Infantry	5th	H	3C		
Ohio	Infantry	51st	H	4?	***	
Ohio	Infantry	61st	H	10A		
Ohio	Infantry	61st	G	3D		
Ohio	Infantry	107th	B	3D		
Ohio	Infantry	107th	D	3D		

State	Type	Unit	Company	Disc style	Source	Rank
Ohio	Infantry	110th	E	1E		
Ohio	Infantry	122nd	H	1E		
Ohio	Infantry	123rd	I	1E		
Ohio	Infantry	126th	G?	1E?	***	
Pennsylvania	Infantry	1st PRVC	F	1B	***	
Pennsylvania	Infantry	2nd PRVC	H	2A		
Pennsylvania	Infantry	3rd PRVC	H	1C		
Pennsylvania	Infantry	3rd PRVC	K	1C	***	
Pennsylvania	Infantry	5th PRVC	A	1A		
Pennsylvania	Infantry	5th PRVC	D	1A		
Pennsylvania	Infantry	5th PRVC	F	1A		
Pennsylvania	Infantry	5th PRVC	G	1A		
Pennsylvania	Infantry	5th PRVC	I	1F		
Pennsylvania	Infantry	5th PRVC	F?	1F		2nd Lt.
Pennsylvania	Infantry	9th PRVC	E	3A		
Pennsylvania	Infantry	10th PRVC	F	1A	***	
Pennsylvania	Infantry	11th PRVC	C	2A		
Pennsylvania	Infantry	11th PRVC	D	4B?	***	
Pennsylvania	Infantry	11th PRVC	?	5B		
Pennsylvania	Infantry	11th PRVC	C	1A		
Pennsylvania	Infantry	11th PRVC	G	1A		
Pennsylvania	Infantry	11th PRVC	D	2A		
Pennsylvania	Infantry	12th PRVC	E	1A		
Pennsylvania	Infantry	12th PRVC	K	1A		corporal
Pennsylvania	Infantry	13th	B	3A		
Pennsylvania	Infantry	13th	K	3A	**	
Pennsylvania	Infantry	13th	L	3A		
Pennsylvania	Infantry	26th	K	3A	***	
Pennsylvania	Infantry	45th	C	1A		
Pennsylvania	Infantry	45th	G	1A	***	
Pennsylvania	Infantry	47th	H	2B	***	
Pennsylvania	Infantry	47th	B	2B	***	
Pennsylvania	Infantry	47th	H	2B	***	
Pennsylvania	Infantry	48th	F	1B		
Pennsylvania	Infantry	49th	A	5A		Sgt.
Pennsylvania	Infantry	49th	I	5A		
Pennsylvania	Infantry	49th	K	2A		
Pennsylvania	Infantry	56th	A	3A		
Pennsylvania	Infantry	56th	D	1D		
Pennsylvania	Infantry	56th	H	3A		
Pennsylvania	Infantry	56th	K	3A		
Pennsylvania	Infantry	57th	C	2A		
Pennsylvania	Infantry	67th	B	1E		
Pennsylvania	Infantry	67th	H	1E		
Pennsylvania	Infantry	69th	K	3D		
Pennsylvania	Infantry	81st	G	1A		
Pennsylvania	Infantry	82nd	A	2A		
Pennsylvania	Infantry	83rd	E	2A		
Pennsylvania	Infantry	83rd	H	2A	***	
Pennsylvania	Infantry	93rd	G	2A	***	
Pennsylvania	Infantry	99th	I	4C		
Pennsylvania	Infantry	99th	D	4C		
Pennsylvania	Infantry	100th	M	1A		
Pennsylvania	Infantry	102nd	K	3A		
Pennsylvania	Infantry	104th	A	3A	***	
Pennsylvania	Infantry	104th	K	3A		
Pennsylvania	Infantry	105th	?	4C		
Pennsylvania	Infantry	115th	I	1A		

State	Type	Unit	Company	Disc style	Source	Rank
Pennsylvania	Infantry	119th	C	2A		
Pennsylvania	Infantry	119th	C	2A		drummer
Pennsylvania	Infantry	119th	A	2A		
Pennsylvania	Infantry	121st	G	2A		
Pennsylvania	Infantry	121st	H	2A		
Pennsylvania	Infantry	121st	B	2A		
Pennsylvania	Infantry	127th	C	3B		
Pennsylvania	Infantry	135th	D	2A	***	
Pennsylvania	Infantry	135th	K	2A		
Pennsylvania	Infantry	137th	A	1F	***	
Pennsylvania	Infantry	137th	B	1F		
Pennsylvania	Infantry	137th	H	1F		
Pennsylvania	Infantry	138th	B	2A		
Pennsylvania	Infantry	142nd	C	2A		
Pennsylvania	Infantry	142nd	I	3D	***	
Pennsylvania	Infantry	148th	C	2D	***	
Pennsylvania	Infantry	148th	C	5A		
Pennsylvania	Infantry	148th	E	5A		
Pennsylvania	Infantry	148th	D	1H	***	
Pennsylvania	Infantry	149th	K	3A		
Pennsylvania	Infantry	149th	G	3B		
Pennsylvania	Infantry	149th PBV	F	3D	***	
Pennsylvania	Infantry	150th	B	3A		
Pennsylvania	Infantry	151st	C	1A	***	
Pennsylvania	Infantry	153rd	G	3A		
Pennsylvania	Infantry	155th	B	1A	***	
Pennsylvania	Infantry	155th	D	2C		
Pennsylvania	Infantry	155th	H	2C		
Pennsylvania	Cavalry	2nd	M	2A		
Pennsylvania	Cavalry	3rd	M	2A		
Pennsylvania	Cavalry	5th	D	9B		
Pennsylvania	Cavalry	6th	H	3C		
Pennsylvania	Cavalry	8th	K	3A		
Pennsylvania	Cavalry	19th	D	1D		
Pennsylvania	Cavalry	19th	L	1D	***	
Pennsylvania	Artillery	Nevin's Ind. Batty.	?	1C		
Pennsylvania	Artillery	Ind.	Batt. D	3A		
Pennsylvania	Artillery	2nd Heavy	Batt. B	1C		
Pennsylvania	Artillery	2nd Heavy	Batt. H	1C		
Pennsylvania	Artillery	2nd Heavy	Batt. M	6A		
Pennsylvania	Artillery	2nd Heavy	Batt. D	1C		
Pennsylvania	Artillery	2nd Heavy	Batt. F	6A		
Rhode Island	Infantry	2nd	G	3C	***	
Rhode Island	Infantry	2nd	K	3C		
Rhode Island	Infantry	2nd	K	3C		Captain
Rhode Island	Infantry	3rd	A	3C	***	
Rhode Island	Infantry	4th	G	1B		
Rhode Island	Infantry	4th	H	1B	***	
Rhode Island	Infantry	4th	B	3C		
Rhode Island	Infantry	4th	C	1B		
Rhode Island	Infantry	4th	C	1B		
Rhode Island	Infantry	4th	C	1B		
Rhode Island	Infantry	4th	K	1B		
Rhode Island	Infantry	7th	I	3C		
Rhode Island	Infantry	11th	B	3C	***	
Rhode Island	Infantry	11th	C	3C		
Rhode Island	Infantry	11th	D	3C	***	

5 • Survey of Identification Discs by State and Style

State	Type	Unit	Company	Disc style	Source	Rank
Rhode Island	Infantry	11th	E	3C	***	
Rhode Island	Infantry	11th	K	3C	***	
Rhode Island	Infantry	11th	B	3C		
Rhode Island	Infantry	12th	E	1A		
Rhode Island	Infantry	12th	D	3C		
Rhode Island	Infantry	12th	G	3C		
Rhode Island	Infantry	12th	I	3C	***	
Rhode Island	?	?	?	3C	***	
Rhode Island	Artillery	1st Light	B	3C	***	
Rhode Island	Artillery	1st Reg.	H	1C	***	
Tenn	Cavalry	9th	L	4B		
Vermont	Infantry	2nd	A	5A	***	
Vermont	Infantry	2nd	B	5A		
Vermont	Infantry	2nd	C	5A	***	
Vermont	Infantry	2nd	H	5A	***	
Vermont	Infantry	2nd	H	5A		
Vermont	Infantry	2nd	H	2A		
Vermont	Infantry	2nd	K	5A	***	
Vermont	Infantry	2nd	K	5A		
Vermont	Infantry	2nd	F	5A		
Vermont	Infantry	2nd	C	5A		
Vermont	Infantry	3rd	A	1B		
Vermont	Infantry	3rd	C	5A		
Vermont	Infantry	3rd	D	5A		
Vermont	Infantry	3rd	E	5A		
Vermont	Infantry	3rd	E	5A		
Vermont	Infantry	3rd	F	5A	***	
Vermont	Infantry	3rd	H	5A		
Vermont	Infantry	3rd	I	5A		
Vermont	Infantry	3rd	I	5A		
Vermont	Infantry	3rd	K	5A		
Vermont	Infantry	3rd	A	6A	***	
Vermont	Infantry	4th	E	5A		
Vermont	Infantry	4th	G	5A		
Vermont	Infantry	4th	G	5A		
Vermont	Infantry	4th	E	4A		
Vermont	Infantry	4th	K	?		
Vermont	Infantry	5th	A	5A	***	
Vermont	Infantry	5th	A	5A	***	
Vermont	Infantry	5th	C	5A		
Vermont	Infantry	5th	D	5A	***	
Vermont	Infantry	5th	D	5A		
Vermont	Infantry	5th	D	5A		
Vermont	Infantry	5th	F	5A		
Vermont	Infantry	5th	G	5A	***	Lt.
Vermont	Infantry	5th	H	5A	***	Lt.
Vermont	Infantry	5th	H	5A		
Vermont	Infantry	5th	I	5A	***	Corporal
Vermont	Infantry	5th	I	5A		
Vermont	Infantry	5th	I	5A		
Vermont	Infantry	5th	I	5A		
Vermont	Infantry	5th	I	5A		
Vermont	Infantry	6th	B	5A		Lt.
Vermont	Infantry	6th	B	5A	***	
Vermont	Infantry	6th	B	5A	***	Captain
Vermont	Infantry	6th	A	5A		
Vermont	Infantry	6th	C	5A		
Vermont	Infantry	6th	E	5A		

State	Type	Unit	Company	Disc style	Source	Rank
Vermont	Infantry	6th	G	5A		Sgt.
Vermont	Infantry	6th	G	5A		
Vermont	Infantry	6th	F	5A		
Vermont	Infantry	6th	K	5A		
Vermont	Infantry	6th	G	4A		
Vermont	Infantry	13th	B	1A		
Vermont	Infantry	13th	B	6A		
Vermont	Infantry	14th	G	5A	***	presentation piece
Vermont	Infantry	14th	G	4C		
Vermont	Infantry	14th	C	1A	***	
Vermont	Infantry	14th	E	1A	***	drummer
Vermont	Infantry	14th	H	1A		
Vermont	Infantry	14th	B	5A		
Vermont	Infantry	15th	B	5A		
Vermont	Infantry	15th	B	6A		
Vermont	Infantry	16th	B	5A	***	
Vermont	Infantry	16th	B	6A		
Vermont	Infantry	16th	C	1A	***	
Vermont	Infantry	16th	I	5A	***	
Vermont	Infantry	17th	H	2B		
Vermont	Infantry	17th	H	2B		
Vermont	Artillery	1st	M	1C		
Vermont	Artillery	1st Heavy/11th	M	1?	**	
Vermont	Artillery	3rd	C	5A		
Vermont	Artillery	11th	D	6A		
Virginia	Infantry	9th	A	1E		
Virginia	Infantry	10th	G	1E		
Virginia	Cavalry	1st	E	1A	***	
Wisconsin	Infantry	2nd	G	3C		
Wisconsin	Infantry	5th	F	5A	***	
Wisconsin	Infantry	5th	H	5A		
Wisconsin	Infantry	6th	E	3A		
Wisconsin	Infantry	6th	E	3A		
Wisconsin	Infantry	6th	K	5A	***	
Wisconsin	Infantry	6th	B	5A?	***	Captain
USA	Infantry	1st Vol.	F	5D		
USA	Infantry	5th CV	H	1D		
USA	Infantry	5th CV	I	1D		
USA	Infantry	5th CV	I	1D		
USA	Infantry	14th	E	1A		
USA	Infantry	14th	F	1A		
USA	Infantry	14th	E	1D		
USA	Infantry	19th VRC	G	6A		
USA	Infantry	1 USSS	E	1B		
USA	Infantry	2 USSS	C	3B		
USA	Cavalry	6th	L	5A		Captain
USA	Artillery	5th	C	1C		

B. Arranged by Style and Substyle

Style	Unit	State	Type	Company	Source	Rank
1A	1st	California	Infantry	B	***	
1A	8th	Connecticut	Infantry	F		
1A	2nd	Maryland	Infantry	F	***	
1A	16th	Massachusetts	Infantry	I		
1A	19th	Massachusetts	Infantry			Sgt. Maj.
1A	19th	Massachusetts	Infantry	B		
1A	21st	Massachusetts	Infantry	G	***	
1A	3rd	Michigan	Infantry	I		
1A	8th	Michigan	Infantry	B		Sutler
1A	8th	New Hampshire	Infantry	E	***	
1A	14th	New Hampshire	Infantry	G		
1A	10th	New Jersey	Infantry	B		
1A	2nd	New York	Artillery	A		
1A	2nd	New York	Artillery	B		
1A	19th	New York	Artillery	D		
1A	43rd	New York	Infantry	H		
1A	108th	New York	Infantry	K	***	
1A	108th	New York	Infantry	K		
1A	5th PRVC	Pennsylvania	Infantry	A		
1A	5th PRVC	Pennsylvania	Infantry	D		
1A	5th PRVC	Pennsylvania	Infantry	F		
1A	5th PRVC	Pennsylvania	Infantry	G		
1A	10th PRVC	Pennsylvania	Infantry	F	***	
1A	11th PRVC	Pennsylvania	Infantry	C		
1A	11th PRVC	Pennsylvania	Infantry	G		
1A	12th PRVC	Pennsylvania	Infantry	K		corporal
1A	12th PRVC	Pennsylvania	Infantry	E		
1A	45th	Pennsylvania	Infantry	C		
1A	45th	Pennsylvania	Infantry	G	***	
1A	81st	Pennsylvania	Infantry	G		
1A	100th	Pennsylvania	Infantry	M		
1A	115th	Pennsylvania	Infantry	I		
1A	151st	Pennsylvania	Infantry	C	***	
1A	155th	Pennsylvania	Infantry	B	***	
1A	12th	Rhode Island	Infantry	E		
1A	14th	USA	Infantry	E		
1A	14th	USA	Infantry	F		
1A	13th	Vermont	Infantry	B		
1A	14th	Vermont	Infantry	C	***	
1A	14th	Vermont	Infantry	E	***	drummer
1A	14th	Vermont	Infantry	H		
1A	16th	Vermont	Infantry	C	***	
1A	107th	New York	Infantry	I	***	
1A	15th	New Hampshire	Infantry	D	***	
1A	21st	New York	Cavalry	A		
1A	1st	Virginia	Cavalry	E		
1B	8th	Connecticut	Infantry	B		
1B	2nd	Delaware	Infantry	C		
1B	5th	New Hampshire	Infantry	A		
1B	1st	New York	Artillery	G		
1B	13th	New York	Infantry	A		Captain
1B	13th NG	New York	Infantry	D		
1B	14th	New York	Infantry	A		
1B	22nd	New York	Infantry	C	***	
1B	22nd	New York	Infantry	D		

Style	Unit	State	Type	Company	Source	Rank
1B	22nd	New York	Infantry	E		
1B	22nd	New York	Infantry	G	***	
1B	57th	New York	Infantry	C		
1B	66th	New York	Infantry	H		
1B	88th	New York	Infantry	H		
1B	1st PRVC	Pennsylvania	Infantry	F	***	
1B	48th	Pennsylvania	Infantry	F		
1B	4th	Rhode Island	Infantry	G		
1B	4th	Rhode Island	Infantry	H	***	
1B	4th	Rhode Island	Infantry	C		
1B	4th	Rhode Island	Infantry	C		
1B	4th	Rhode Island	Infantry	K		
1B	3rd	Vermont	Infantry	A		
1B	1 USSS	USA	Infantry	E		
1C	1st H. Art.	Connecticut	Artillery	D		
1C	1st H. Art.	Connecticut	Artillery	D		
1C	27th	Maine	Infantry	G		
1C	4th	Michigan	Infantry	K	***	
1C	13th	New Hampshire	Infantry	E		
1C	1st	New Jersey	Infantry	F		
1C	21st	New Jersey	Infantry	E	***	
1C	145th	New York	Infantry	I		
1C	1st vet.	New York	Cavalry	C	***	
1C	1st vet.	New York	Cavalry	I	***	presentation to lady
1C	1st vet.	New York	Cavalry	M		
1C	15th	New York	Cavalry	K		
1C	18th	New York	Cavalry	D		
1C	18th	New York	Cavalry	E		
1C	18th	New York	Cavalry	G		
1C	18th	New York	Cavalry	G		
1C	21st	New York	Cavalry	C	***	
1C	21st	New York	Cavalry	H		
1C	21st Vet.	New York	Cavalry	A		Corporal
1C	24th	New York	Cavalry	H		
1C	31st	New York	Infantry	A		
1C	43rd	New York	Infantry	D		
1C	51st	New York	Infantry	B		
1C	63rd	New York	Infantry	A		
1C	153rd	New York	Infantry	D		To My Mother
1C	153rd	New York	Infantry	E		
1C	153rd	New York	Infantry	I		
1C	157th	New York	Infantry	A		
1C	2nd H. Art.	Pennsylvania	Artillery	Batt. B		
1C	2nd H. Art.	Pennsylvania	Artillery	Batt. H		
1C	2nd H. Art.	Pennsylvania	Artillery	Batt. D		
1C	3rd PRVC	Pennsylvania	Infantry	H		
1C	3rd PRVC	Pennsylvania	Infantry	K	***	
1C	Nevin's Ind Batty	Pennsylvania	Artillery			
1C	1st Reg	Rhode Island	Artillery	H	***	
1C	5th	USA	Artillery	C		
1C	1st	Vermont	Artillery	M		
1D	2nd	Delaware	Infantry	G		
1D	7th	Massachusetts	Infantry	C		
1D	1st Vet.	New York	Cavalry	B		
1D	2nd	New York	Cavalry	E		
1D	2nd H. Art.	New York	Artillery	M		

Style	Unit	State	Type	Company	Source	Rank
1D	2nd Vet.	New York	Cavalry	C		
1D	20th	New York	Infantry	B		
1D	20th	New York	Cavalry	B		
1D	19th	Pennsylvania	Cavalry	D		
1D	19th	Pennsylvania	Cavalry	L	***	
1D	56th	Pennsylvania	Infantry	D		
1D	14th	USA	Infantry	E		
1D	5th CV	USA	Infantry	H		
1D	5th CV	USA	Infantry	I		
1D	5th CV	USA	Infantry	I		
1D	23rd	Massachusetts	Infantry	G		
1E	110th	Ohio	Infantry	E		
1E	122nd	Ohio	Infantry	H		
1E	123rd	Ohio	Infantry	I		
1E	67th	Pennsylvania	Infantry	B		
1E	67th	Pennsylvania	Infantry	H		
1E	9th	Virginia	Infantry	A		
1E	10th	Virginia	Infantry	G		
1E	126th	Ohio	Infantry	G?	***	
1F	6th	New York	Artillery	L		
1F	51st	New York	Infantry	C		
1F	5th	Pennsylvania	Infantry	F		2nd Lt.
1F	5th PRVC	Pennsylvania	Infantry	I		
1F	137th	Pennsylvania	Infantry	A	***	
1F	137th	Pennsylvania	Infantry	B		
1F	137th	Pennsylvania	Infantry	H		
1H	141st	New York	Infantry	K		
1H	148th	Pennsylvania	Infantry	D	***	
1H	28th	Massachusetts	Infantry	?	***	
1L	3rd	Maine	Infantry	E		
1L	3rd	Maine	Infantry	E		
2A	11th	Connecticut	Infantry	G		
2A	1st	Delaware	Infantry	I		
2A	7th	Maryland	Infantry	D	***	
2A	9th	Massachusetts	Infantry	B		Sgt.
2A	32nd	Massachusetts	Infantry	A	***	
2A	2nd	New York	Cavalry	G		
2A	2nd	New York	Cavalry	B		
2A	2nd sm (82nd)	New York	Infantry	I	***	
2A	9th	New York	Cavalry	H	***	
2A	14th Brklyn.	New York	Infantry	D	***	
2A	14th Brklyn.	New York	Infantry	E	***	
2A	30th	New York	Infantry	G	***	Asst. Surg.
2A	30th	New York	Infantry	G		
2A	30th	New York	Infantry	E		Sgt.
2A	34th	New York	Infantry	D		Sgt.
2A	34th	New York	Infantry	K		
2A	50th Eng.	New York	Infantry	B		
2A	69th	New York	Infantry	E	***	
2A	82nd	New York	Infantry	I	***	
2A	121st	New York	Infantry	I		
2A	140th	New York	Infantry	G		
2A	140th	New York	Infantry	H	***	
2A	164th	New York	Infantry	C		
2A	169th	New York	Infantry	F	***	Sgt.
2A	169th	New York	Infantry	G		Drummer
2A	169th	New York	Infantry	K		Corporal

Style	Unit	State	Type	Company	Source	Rank
2A	2nd	Pennsylvania	Cavalry	M		
2A	2nd PRVC	Pennsylvania	Infantry	H		
2A	3rd	Pennsylvania	Cavalry	M		
2A	11 PRVC	Pennsylvania	Infantry	D		
2A	11th PRVC	Pennsylvania	Infantry	C		
2A	49th	Pennsylvania	Infantry	K		
2A	57th	Pennsylvania	Infantry	C		
2A	82nd	Pennsylvania	Infantry	A		
2A	83rd	Pennsylvania	Infantry	E		
2A	83rd	Pennsylvania	Infantry	H	***	
2A	93rd	Pennsylvania	Infantry	G	***	
2A	119th	Pennsylvania	Infantry	C		
2A	119th	Pennsylvania	Infantry	C		Drummer
2A	119th	Pennsylvania	Infantry	A		
2A	121st	Pennsylvania	Infantry	G		
2A	121st	Pennsylvania	Infantry	H		
2A	121st	Pennsylvania	Infantry	B		
2A	135th	Pennsylvania	Infantry	D	***	
2A	135th	Pennsylvania	Infantry	K		
2A	138th	Pennsylvania	Infantry	B		
2A	142nd	Pennsylvania	Infantry	C		
2A	2nd	Vermont	Infantry	H		
2A	153rd	New York	Infantry	A		
2B	53rd	Massachusetts	Infantry	K		
2B	58th	Massachusetts	Infantry	A	***	
2B	1st	Michigan	Infantry	E		
2B	96th	New York	Infantry	C		Sgt.
2B	156th	New York	Infantry	G	***	
2B	47th	Pennsylvania	Infantry	H	***	
2B	47th	Pennsylvania	Infantry	B	***	
2B	47th	Pennsylvania	Infantry	H	***	
2B	17th	Vermont	Infantry	H		
2B	17th	Vermont	Infantry	H		
2C	10th	Maryland	Infantry	A		
2C	10th	Maryland	Infantry	A		
2C	1st Md PHB	Maryland	Infantry	I		
2C	17th	New York	Infantry	H		
2C	17th	New York	Infantry	H		
2C	59th	New York	Infantry	I	***	
2C	95th	New York	Infantry	B		
2C	97th	New York	Infantry	D		
2C	147th	New York	Infantry	E		
2C	155th	Pennsylvania	Infantry	D		
2C	155th	Pennsylvania	Infantry	H		
2C	9th	Massachusetts	Infantry	G		
2C	159th	New York	Infantry	H	***	Presented to
2D	9th	New Hampshire	Infantry	B		
2D	148th	Pennsylvania	Infantry	C	***	
3A	2nd	D.C.	Infantry	E		
3A	20th	Indiana	Infantry	C	***	
3A	3rd	Indiana	Cavalry	B		
3A	3rd	Maine	Infantry	A		
3A	1st	Massachusetts	Infantry	H		
3A	1st	Massachusetts	Cavalry	K		
3A	4th	Massachusetts	Cavalry	G	***	
3A	10th	Massachusetts	Infantry	D		
3A	35th	Massachusetts	Infantry	E	***	
3A	42nd	Massachusetts	Infantry	I	***	

Style	Unit	State	Type	Company	Source	Rank
3A	44th	Massachusetts	Infantry	I		
3A	56th	Massachusetts	Infantry	H	***	
3A	59th	Massachusetts	Infantry	I	***	
3A	2nd	Michigan	Infantry	C	***	Drummer (dug)
3A	2nd	Michigan	Infantry	I		
3A	2nd	New Hampshire	Infantry	C		
3A	2nd	New Hampshire	Infantry	H		
3A	2nd	New Hampshire	Infantry	E		
3A	6th	New Jersey	Infantry	A		
3A	6th	New Jersey	Infantry	G		
3A	2nd	New York	Infantry	C		
3A	2nd	New York	Infantry	E		Sgt.
3A	2nd	New York	Infantry	I		
3A	2nd FZ	New York	Cavalry	K		
3A	6th	New York	Infantry	A		
3A	21st	New York	Infantry			
3A	52nd	New York	Infantry	G		
3A	70th	New York	Infantry	D		
3A	71st	New York	Infantry	C		
3A	72nd	New York	Infantry	E	***	
3A	74th	New York	Infantry	I		Sgt.
3A	74th	New York	Infantry	I		
3A	77th	New York	Infantry	C		
3A	87th	New York	Infantry	C		
3A	87th	New York	Infantry	I		
3A	Harris Light (2nd)	New York	Cavalry	F		
3A	Harris Light (2nd)	New York	Cavalry	L		
3A	8th	Pennsylvania	Cavalry	K		
3A	9th PRVC	Pennsylvania	Infantry	E		
3A	13th	Pennsylvania	Infantry	B		
3A	13th	Pennsylvania	Infantry	K	**	
3A	13th	Pennsylvania	Infantry	L		
3A	26th	Pennsylvania	Infantry	K	***	
3A	56th	Pennsylvania	Infantry	A		
3A	56th	Pennsylvania	Infantry	H		
3A	56th	Pennsylvania	Infantry	K		
3A	102nd	Pennsylvania	Infantry	K		
3A	104th	Pennsylvania	Infantry	A	***	
3A	104th	Pennsylvania	Infantry	K		
3A	149th	Pennsylvania	Infantry	K		
3A	150th	Pennsylvania	Infantry	B		
3A	153rd	Pennsylvania	Infantry	G		
3A	Ind Pa. Art.	Pennsylvania	Artillery	Batt. D		
3A	6th	Wisconsin	Infantry	E		
3B	8th	Connecticut	Infantry	G		
3B	12th	Connecticut	Infantry	K		
3B	21st	Connecticut	Infantry	C		
3B	101st	New York	Infantry	A	***	Drummer
3B	169th	New York	Infantry	H		
3B	127th	Pennsylvania	Infantry	C		
3B	149th	Pennsylvania	Infantry	G		
3B	9th Batty	Massachusetts	Artillery	A		
3B	11th	New York	Infantry	K		
3B	2nd USSS	USA	Infantry	C		
3C	14th	Connecticut	Infantry	H		

Style	Unit	State	Type	Company	Source	Rank
3C	18th	Connecticut	Infantry	E		
3C	18th	Connecticut	Infantry	B		
3C	21st	Connecticut	Infantry	F		
3C	1st	Massachusetts	Infantry	C		
3C	3rd	Massachusetts	Infantry	C		
3C	3rd	Massachusetts	Infantry	C		Captain
3C	6th	New Jersey	Infantry	G		
3C	33rd	New Jersey	Infantry	D		
3C	1st light	New York	Artillery	D	***	
3C	101st	New York	Infantry	F		
3C	145th	New York	Infantry	D		
3C	Rock Batt	New York	Artillery	B		
3C	5th	Ohio	Infantry	H		
3C	6th	Pennsylvania	Cavalry	H		
3C	?	Rhode Island		?	***	
3C	1st light	Rhode Island	Artillery	B	***	
3C	2nd	Rhode Island	Infantry	G	***	
3C	2nd	Rhode Island	Infantry	K		
3C	2nd	Rhode Island	Infantry	K		Captain
3C	3rd	Rhode Island	Infantry	A	***	
3C	4th	Rhode Island	Infantry	B		
3C	7th	Rhode Island	Infantry	I		
3C	11th	Rhode Island	Infantry	B	***	
3C	11th	Rhode Island	Infantry	C		
3C	11th	Rhode Island	Infantry	D	***	
3C	11th	Rhode Island	Infantry	E	***	
3C	11th	Rhode Island	Infantry	K	***	
3C	11th	Rhode Island	Infantry	B		
3C	12th	Rhode Island	Infantry	D		
3C	12th	Rhode Island	Infantry	G		
3C	12th	Rhode Island	Infantry	I	***	
3C	2nd	Wisconsin	Infantry	G		
3D	18th	Connecticut	Infantry	E		
3D	19th	Maine	Infantry	G		
3D	1st	New Jersey	Infantry	F		
3D	61st	Ohio	Infantry	G		
3D	107th	Ohio	Infantry	B		
3D	107th	Ohio	Infantry	D	***	
3D	69th	Pennsylvania	Infantry	K		
3D	142nd	Pennsylvania	Infantry	I	***	
3D	149th PBV	Pennsylvania	Infantry	F	***	
4A	18th	Michigan	Infantry	F		
4A	4th	Vermont	Infantry	E		
4A	6th	Vermont	Infantry	G		
4B	19th	Connecticut	Infantry	E		
4B	17th	Indiana	Infantry	B		
4B	20th	Indiana	Infantry	H		
4B	147th	Indiana	Infantry	B		
4B	17th	Maine	Infantry	B		
4B	34th	Massachusetts	Infantry	A	***	
4B	11th	New Jersey	Infantry	D	***	
4B	3rd Excelsior	New York	Infantry	D		
4B	5th Excelsior	New York	Infantry	D		
4B	15th Eng.	New York	Infantry	L		
4B	25th	New York	Infantry	C	***	
4B	25th	New York	Infantry	D	***	
4B	37th	New York	Infantry	I	***	
4B	74th	New York	Infantry	C		

5 • Survey of Identification Discs by State and Style

Style	Unit	State	Type	Company	Source	Rank
4B	9th	Tennessee	Cavalry	L		
4B	72nd	Pennsylvania	Infantry	D		
4B	74th	Pennsylvania	Infantry	D		
4B	3rd	New York	Artillery	E		
4B	6th Ind. Batty	New York	Artillery	?		
4B	1st MR	New York	Cavalry	K		
4B	1st vet	New York	Cavalry	C		
4B	1st. MD. PHB	Maryland	Infantry	A		
4B	1st H. Art.	Maine	Artillery	H		
4B	15th	New Hampshire	Infantry	C		
4B	11th	Pennsylvania	Infantry	D	***	
4B	4th	Vermont	Infantry	K		
4C	9th	Indiana	Infantry	C	***	
4C	20th	Indiana	Infantry	H		
4C	1st	Massachusetts	Infantry	K		
4C	16th	Massachusetts	Infantry	K		
4C	20th	Massachusetts	Infantry	?		
4C	3rd	Michigan	Infantry	A		
4C	2nd	New Hampshire	Infantry	F		
4C	5th	New Jersey	Infantry	E		
4C	7th	New Jersey	Infantry	H		
4C	5th	New York	Infantry	G		
4C	25th	New York	Infantry			
4C	40th	New York	Infantry	E		
4C	120th	New York	Infantry	C		
4C	99th	Pennsylvania	Infantry	I		
4C	99th	Pennsylvania	Infantry	D		
4C	105th	Pennsylvania	Infantry	?		
4C	14th	Vermont	Infantry	G		
5A	2nd	Delaware	Infantry	B		
5A	2nd	Delaware	Infantry	E		
5A	2nd	Delaware	Infantry	H	***	
5A	2nd	Delaware	Infantry	H		
5A	58th	Illinois	Infantry	?		Sgt.
5A	7th	Maine	Infantry	F	***	
5A	7th	Maine	Infantry	H	***	
5A	2nd	Massachusetts	Infantry	G	***	
5A	12th	Massachusetts	Infantry	B	***	
5A	28th	Massachusetts	Infantry	H	***	
5A	8th	Michigan	Infantry			sutler
5A	1st	New Hampshire	Cavalry	L		
5A	1st	New Hampshire	Artillery	?		
5A	1st	New Hampshire	Artillery	?		
5A	2nd	New Hampshire	Infantry	?		
5A	5th	New Hampshire	Infantry	B	***	
5A	5th	New Hampshire	Infantry	B		
5A	5th	New Hampshire	Infantry	D		Corporal
5A	5th	New Hampshire	Infantry	F	***	
5A	5th	New Hampshire	Infantry	I	***	
5A	6th	New Hampshire	Infantry	B	***	
5A	6th	New Hampshire	Infantry	I		
5A	9th	New Hampshire	Infantry	B	***	
5A	9th	New Hampshire	Infantry	C	***	
5A	9th	New Hampshire	Infantry	C	***	
5A	9th	New Hampshire	Infantry	D	***	
5A	9th	New Hampshire	Infantry	D		
5A	9th	New Hampshire	Infantry	F		

Style	Unit	State	Type	Company	Source	Rank
5A	10th	New Hampshire	Infantry	A		Sgt.
5A	10th	New Hampshire	Infantry	B		
5A	10th	New Hampshire	Infantry	D		
5A	10th	New Hampshire	Infantry	H		
5A	10th	New Hampshire	Infantry	H		
5A	10th	New Hampshire	Infantry	H		Sgt.
5A	10th	New Hampshire	Infantry	H		
5A	10th	New Hampshire	Infantry	H		Sgt.
5A	10th	New Hampshire	Infantry	I		
5A	11th	New Hampshire	Infantry	C		
5A	11th	New Hampshire	Infantry	D	***	
5A	11th	New Hampshire	Infantry	E		
5A	11th	New Hampshire	Infantry	E		Cook
5A	11th	New Hampshire	Infantry	B		Cook
5A	11th	New Hampshire	Infantry	E		
5A	11th	New Hampshire	Infantry	H		
5A	11th	New Hampshire	Infantry	I		
5A	12th	New Hampshire	Infantry	B		
5A	12th	New Hampshire	Infantry	D		
5A	12th	New Hampshire	Infantry	D		
5A	12th	New Hampshire	Infantry	F		
5A	12th	New Hampshire	Infantry	G		
5A	12th	New Hampshire	Infantry	G		
5A	12th	New Hampshire	Infantry	I?	***	
5A	13th	New Hampshire	Infantry	B		
5A	13th	New Hampshire	Infantry	C	***	Corporal
5A	13th	New Hampshire	Infantry	D		
5A	13th	New Hampshire	Infantry	H		
5A	13th	New Hampshire	Infantry	K		
5A	14th	New Hampshire	Infantry	A		
5A	14th	New Hampshire	Infantry	B		
5A	14th	New Hampshire	Infantry	C		
5A	14th	New Hampshire	Infantry	D		
5A	14th	New Hampshire	Infantry	F		
5A	14th	New Hampshire	Infantry	F	***	
5A	14th	New Hampshire	Infantry	H		Corporal
5A	14th	New Hampshire	Infantry	H		
5A	14th	New Hampshire	Infantry	H?		
5A	14th	New Hampshire	Infantry	K		
5A	15th	New Hampshire	Infantry	A	***	Sgt.
5A	15th	New Hampshire	Infantry	B	***	
5A	15th	New Hampshire	Infantry	C	***	
5A	15th	New Hampshire	Infantry	C		
5A	15th	New Hampshire	Infantry	G	***	
5A	15th	New Hampshire	Infantry	I	***	
5A	15th	New Hampshire	Infantry	I	***	
5A	15th	New Hampshire	Infantry	I	***	
5A	15th	New Hampshire	Infantry	K	***	
5A	15th	New Hampshire	Infantry	K		
5A	15th	New Hampshire	Infantry	H	***	
5A	15th	New Hampshire	Infantry	G	***	
5A	15th	New Hampshire	Infantry	I		
5A	15th	New Hampshire	Infantry	I		
5A	15th	New Hampshire	Infantry	D		
5A	14th	New Jersey	Infantry	A	***	
5A	1st	New York	Infantry	I		
5A	1st	New York	Artillery	L		
5A	1st	New York	Artillery	C		

Style	Unit	State	Type	Company	Source	Rank
5A	7th	New York	Infantry	K	***	
5A	31st	New York	Infantry	A	***	
5A	31st	New York	Infantry	A		
5A	33rd	New York	Infantry	B		
5A	33rd	New York	Infantry	B		
5A	33rd	New York	Infantry	G	***	
5A	57th	New York	Infantry	F		
5A	57th	New York	Infantry	H		Lt.
5A	57th	New York	Infantry	H		Sgt.
5A	61st	New York	Infantry	A	***	
5A	61st	New York	Infantry	F	***	
5A	61st	New York	Infantry	F		
5A	63rd	New York	Infantry	A		
5A	63rd	New York	Infantry	F		
5A	64th	New York	Infantry	G		
5A	69th	New York	Infantry	B	***	
5A	69th	New York	Infantry	F	***	
5A	69th	New York	Infantry	I	***	
5A	76th	New York	Infantry	A	***	
5A	76th	New York	Infantry	E		
5A	76th	New York	Infantry	H		
5A	76th	New York	Infantry	I	***	
5A	49th	Pennsylvania	Infantry	A		Sgt.
5A	49th	Pennsylvania	Infantry	I		
5A	148th	Pennsylvania	Infantry	C		
5A	148th	Pennsylvania	Infantry	E		
5A	6th	USA	Cavalry	L		Captain
5A	2nd	Vermont	Infantry	A	***	
5A	2nd	Vermont	Infantry	B		
5A	2nd	Vermont	Infantry	C	***	
5A	2nd	Vermont	Infantry	H	***	
5A	2nd	Vermont	Infantry	H		
5A	2nd	Vermont	Infantry	K	***	
5A	2nd	Vermont	Infantry	K		
5A	2nd	Vermont	Infantry	F		
5A	2nd	Vermont	Infantry	C		
5A	3rd	Vermont	Infantry	C		
5A	3rd	Vermont	Infantry	D		
5A	3rd	Vermont	Infantry	E		
5A	3rd	Vermont	Infantry	E		
5A	3rd	Vermont	Infantry	F	***	
5A	3rd	Vermont	Infantry	H		
5A	3rd	Vermont	Infantry	I		
5A	3rd	Vermont	Infantry	I		
5A	3rd	Vermont	Infantry	K		
5A	3rd	Vermont	Artillery	C		
5A	4th	Vermont	Infantry	E		
5A	4th	Vermont	Infantry	G		
5A	4th	Vermont	Infantry	G		
5A	5th	Vermont	Infantry	A	***	
5A	5th	Vermont	Infantry	A	***	
5A	5th	Vermont	Infantry	C		
5A	5th	Vermont	Infantry	D	***	
5A	5th	Vermont	Infantry	D		
5A	5th	Vermont	Infantry	D		
5A	5th	Vermont	Infantry	F		
5A	5th	Vermont	Infantry	G	***	Lt.
5A	5th	Vermont	Infantry	H	***	Lt.

Style	Unit	State	Type	Company	Source	Rank
5A	5th	Vermont	Infantry	H		
5A	5th	Vermont	Infantry	I	***	Corporal
5A	5th	Vermont	Infantry	I		
5A	5th	Vermont	Infantry	I		
5A	5th	Vermont	Infantry	I		
5A	6th	Vermont	Infantry	B		Lt.
5A	6th	Vermont	Infantry	B	***	
5A	6th	Vermont	Infantry	B	***	Captain
5A	6th	Vermont	Infantry	A		
5A	6th	Vermont	Infantry	C		
5A	6th	Vermont	Infantry	E		
5A	6th	Vermont	Infantry	G		Sgt.
5A	6th	Vermont	Infantry	G		
5A	6th	Vermont	Infantry	K		
5A	14th	Vermont	Infantry	G	***	presentation piece
5A	14th	Vermont	Infantry	B		
5A	15th	Vermont	Infantry	B		
5A	16th	Vermont	Infantry	B	***	
5A	16th	Vermont	Infantry	I	***	
5A	5th	Wisconsin	Infantry	F	***	
5A	5th	Wisconsin	Infantry	H		
5A	6th	Wisconsin	Infantry	K	***	
5A	14th	New Hampshire	Infantry	C	***	Corporal
5A	14th	New Hampshire	Infantry	I	***	
5A	6th	Wisconsin	Infantry	B	***	Captain
5B	26th	Michigan	Infantry	A		
5B	15th	New Hampshire	Infantry	D	***	
5B	13th H. Art.	New York	Artillery	A		
5B	13th H. Art.	New York	Artillery	A		
5B	11th	Pennsylvania	Infantry	?		
5B	5th	New Hampshire	Infantry	H	***	
5B	5th	New Hampshire	Infantry	H	***	
5B	89th	New York	Infantry	K		
5C	8th	Connecticut	Infantry	K		
5C	11th	Connecticut	Infantry	D		
5C	11th	Connecticut	Infantry	D		
5C	11th	Connecticut	Infantry	none		
5C	15th	Connecticut	Infantry	G		
5C	11th	New Hampshire	Infantry	F		
5C	11th	New Hampshire	Infantry	I		
5C	3rd	New York	Cavalry	G	***	
5C	3rd	New York	Cavalry	F	***	
5C	10th	New York	Cavalry	L	***	
5C	13th H. Art.	New York	Artillery	D		
5D	16th H. Art.	New York	Artillery	M	***	
5D	99th	New York	Infantry	C	***	
5D	99th	New York	Infantry	F		
5D	1st Vol.	USA	Infantry	F		
6A	12th	Illinois	Cavalry	A	***	
6A	1st	Maryland	Cavalry	G		
6A	8th	Massachusetts	Infantry	E	***	
6A	32nd	Massachusetts	Infantry	H		
6A	41st	Massachusetts	Infantry	A		Bugler
6A	41st	Massachusetts	Infantry	E	***	Lt.
6A	6th	Michigan	Infantry	C		
6A	6th	Michigan	Cavalry	C		
6A	6th	Michigan	Cavalry	C		

Style	Unit	State	Type	Company	Source	Rank
6A	6th	Michigan	Cavalry	H	***	
6A	4th	New Hampshire	Infantry	G	***	
6A	1st	New York	Cavalry	H		
6A	2nd	New York	Infantry	G		
6A	2nd	New York	Cavalry	A		
6A	2nd Mounted Rifles	New York	Cavalry	L	***	
6A	9th	New York	Infantry	A	***	
6A	10th	New York	Cavalry	A	***	
6A	104th	New York	Infantry	G	***	
6A	107th	New York	Infantry	H	***	
6A	124th	New York	Infantry	H	***	Sgt.
6A	2nd H. Art.	Pennsylvania	Artillery	Batt. M		
6A	2nd H. Art.	Pennsylvania	Artillery	Batt. F		
6A	19th VRC	USA	Infantry	G		
6A	3rd	Vermont	Infantry	A	***	
6A	11th	Vermont	Artillery	D		
6A	13th	Vermont	Infantry	B		
6A	15th	Vermont	Infantry	B		
6A	16th	Vermont	Infantry	B		
6A (silver)	59th	Illinois	Infantry	?		
8A	56th	New York	Infantry	K		
8A	80th	New York	Infantry	?		
9A	130th	Indiana	Infantry	I	***	
9A	4th H. Art.	Massachusetts	Artillery	M		
9B	5th	Pennsylvania	Cavalry	D		
10A	2nd	D.C.	Infantry	C		
10A	4th	New York	Cavalry	H		
10A	6th	New York	Infantry	?		
10A	8th SS	New York	Infantry	?		
10A	147th	New York	Infantry	?		
10A	NYSS	New York	Infantry	?		
10A	NYSS	New York	Infantry	?		
10A	61st	Ohio	Infantry	H		
11A	7th	New Jersey	Infantry	none	***	
12A	6th	New Hampshire	Infantry	I		
13A	50th Eng.	New York	Infantry	?		
13A	50th Eng.	New York	Infantry	C		Drummer
13A	165th	New York	Infantry	C		
14A	Excelsior	New York	Infantry	?		
16A	20th	Indiana	Infantry	F		
16A	20th	Indiana	Infantry	?	***	
16A	1st Batt. Vet. Vol.	New Hampshire	Artillery	?		
16A	2nd	New York	Infantry	G		
16A	4th H. Art.	New York	Artillery	H	***	
16A	16th H. Art.	New York	Artillery	A		
16A	125th	New York	Infantry	H	***	

Epilogue

One of the more famous landmarks in the Civil War is Burnside's Bridge on the battlefield of Antietam. It became famous on September 17, 1862, when Major General Ambrose Burnside was ordered to cross Antietam Creek at the bridge that would later bear his name with his 9th Corps and attack the Confederate right flank. One of the brigades that would make an unsuccessful attempt to cross the bridge was Colonel James Nagle's 1st Brigade of the 2nd Division consisting of four regiments from three states; the 2nd Maryland Infantry, the 6th New Hampshire Infantry, the 9th New Hampshire Infantry and the 48th Pennsylvania Infantry.[1] Three of the regiments were veteran units having been recruited in the summer of 1861 and brigaded together since July 22, 1862. On September 6, 1862, the newly recruited 9th New Hampshire joined the brigade in Washington, D.C. The brigade entered the fighting on September 17 after having been engaged at South Mountain on September 14. Who were these men?

The 2nd Maryland Infantry was organized in Baltimore in the late summer and fall of 1861. The regiment was mustered into the Union service in September 1861 to serve for three years. It spent its early service in the Army of the Potomac and was assigned to the 9th Corps in March 1862. It was assigned to the 1st Brigade on July 22, 1862. After that assignment, the brigade was in the Battles of 2nd Bull Run on August 28–31, Chantilly on September 1 and South Mountain on September 14 before it arrived at Antietam.[2]

The 6th New Hampshire Infantry was recruited across the state in the fall of 1861, mustering in between November 27 and November 30, 1861, at Keene, New Hampshire, for three years. After spending the fall of 1861 in Maryland, the 6th joined General Ambrose Burnside's troops in January 1862. It participated in the actions of General Burnside's forces in North Carolina until the 9th Corps returned to Virginia for the 2nd Bull Run Campaign in August 1862. The regiment was in the fighting at the Railroad Cut, and later, on September 1 at the Battle of Chantilly. From there the troops marched to South Mountain and on to the Battle of Antietam.[3]

The 48th Pennsylvania Infantry was recruited in Schuylkill County, Pennsylvania, in the summer of 1861, mustering into the Union Army in September 1861 for three years. It spent the fall of 1861 in North Carolina and moved with the rest of the 9th Corps to Virginia in July 1862 and was present for the 2nd Battle of Bull Run, Chantilly and South Mountain.[4]

The 9th New Hampshire was recruited from all across the state of New Hampshire in the late summer of 1862, for three years. On September 6, 1862, the regiment was assigned to Colonel Nagle's brigade while in the defenses of Washington. It would see its first action on September 14, 1862 just eight days after joining the brigade at the battle of South Mountain where it attacked a Confederate Brigade.[5]

On September 17, 1862, the 9th Corps was on the Union left flank. Around noon it was ordered to cross Antietam Creek at what was then known as the Rohrback Bridge, but would soon bear another name. Shortly thereafter Colonel Nagle's brigade made their unsuccessful attempt to cross the bridge.

Woodcut of Burnside's Bridge at the Antietam battlefield taken from the October 18, 1862, issue of *Harper's Weekly*.

Colonel Nagle would write in his official report of the actions of his the brigade on September 17, 1862, the following:

> On the morning of September 17 my brigade received orders, while encamped near Sharpsburg, to advance on the enemy at a point he had selected where the stone bridge crosses the Antietam Creek, about 2 miles from Sharpsburg. The position was a strong one for the enemy, as he was posted in strong force on the bank of Antietam Creek, on the wooded banks of this stream, with precipitous banks that afforded them shelter from our artillery and infantry. Two roads diverge from this bridge, and the approach to it is through a narrow ravine, admitting not more than one regiment at a time, upon which a deadly volley could be easily poured from the enemy in ambuscade on the other side of the bridge. The topography being of such a nature that the whole brigade could not be posted to advantage, a front was selected on the left, on the banks of the stream, from which an oblique range upon the bridge could be had, while my right occupied the bluff overlooking the bridge, behind which two regiments of the enemy were concealed in the heavy underbrush.
>
> From this strong position the enemy poured a terrific fire upon our infantry, which was replied to in a very spirited manner by all the regiments in my brigade. The Forty-eighth Pennsylvania Volunteers, Lieutenant-Colonel Sigfried Commanding, passed through a field skirted by a forest, in which the enemy was posted, and with the other regiments soon cleared it of the rebel sharpshooters, placed there in concealed positions. The Ninth New Hampshire Volunteers, Col. E.Q. Fellows, was placed near the

Opposite, top: Modern view of Burnside's Bridge from the Confederate positions. *Bottom:* Modern view looking across Burnside's Bridge toward the Confederate positions.

bridge, and opened a destructive fire directly upon the enemy, and expended nearly all their ammunition during a gallant resistance of an hour, in which they were between the fires of two regiments of the enemy, and sustained themselves nobly. Lieut. H.B. Titus, of Ninth New Hampshire Volunteers, fell, badly wounded, at this point, with several of the commissioned officers of the regiment. The Second Maryland and Sixth New Hampshire Volunteers were placed in a perilous position near the bridge, and are entitled to commendation for their soldier-like bearing and bravery displayed.

When most of the ammunition in my brigade was expended, my brigade fell back; the Second Brigade was ordered up to storm the bridge, which they did, and my brigade ordered to follow for the possession of the heights on the opposite side of the river. With cheers and great enthusiasm my brigade passed the bridge and planted our flag on the heights in a few moments. After other re-enforcements arrived we advanced and drove the enemy from their position on the range of hills near the river, where a severe battle took place, in which my brigade maintained their ground, though they were somewhat cut up in this position by the grape and shell of the enemy. Skirmishing was kept up on the 18th between our heavy picket force and the skirmishers of the enemy....

The loss in my brigade on the 17th and 18th was 35 killed, 154 wounded, 15 missing; total, 204.

Enclosed please find the names of all killed, wounded, and missing, and the regiments to which they belong.

All of which is respectfully submitted by your obedient servant,

James Nagle,
Brigadier-General, First Brig., Second Div., Ninth Army Corps."[6]

After the Battle of Antietam, Nagle's Brigade returned to Virginia and was engaged at the Battle of Fredericksburg in December 1862 where it suffered 31 killed, 421 wounded and 48 missing or captured, for a total of 500. However, 266 of the losses were suffered by two other regiments that had joined the brigade, the 7th and 12th Rhode Island.[7] After the battle, the brigade stayed in the vicinity of Fredericksburg until February 1863, when it was sent west with the rest of the 9th Corps. Upon its arrival in Kentucky the brigade was broken up. The 6th and 9th New Hampshire remained in the 2nd Division and were sent to Vicksburg.[8] The 2nd Maryland and 48th Pennsylvania participated in the Knoxville campaign in 1863.[9] After this service the 9th Corps returned to the Eastern Theater and all the regiments, albeit now in two different brigades, eventually saw combat at Spotsylvania, North Anna River, Cold Harbor, Petersburg and Appomattox.

At least four of Nagel's men bought identification discs sometime during their service, and perhaps each disc hung around their necks on that awful day at the Antietam Creek. Those four soldiers were Private Thomas H. Tracy in Company F of the 2nd Maryland, Private James Kirby of Company I of the 6th New Hampshire, Ordnance Sergeant J.W. Jenkins of Company F of the 48th Pennsylvania and Private Charles L. Perry of Company D of the 9th New Hampshire. Each would have a story that would end in different places and times.

The Second Maryland's Thomas Tracy was the first of the four to join the army. Private Tracy's service records (records show his name as both Tracy and Tracey) show that he mustered into Company F on August 8, 1861, for 3 years and he gave his age as 22.[10] The company's descriptive book page says Private Tracy was 5' 9" tall, with Hazel eyes, black hair and with a dark complexion. He is shown as a laborer before enlisting. The records, as was common early in the war, showed his status as "Not stated" from August to December 1861. After that he was listed as "present" from January 1862 to April 1863. During this period the 2nd Maryland was heavily engaged at the Second Battle of Bull Run on August 29, 1862, at Groveton. About 4 P.M. the regiment was ordered to attack across the unfinished railroad in their front. John Hennessey's *Return to Bull Run*, says the "...Federals had advanced only a hundred yards beyond the unfinished railroad when the fire from (Colonel Bradley T.) Johnson's Virginians began to lash at their left and rear."[11] The account continues with, "the 48th (Pennsylvania Volunteer Infantry) [was] flanked to the right along the unfinished railroad, away from the danger. The 2nd Maryland, on the right of the 48th, followed suit." In the fighting during the entire campaign the

2nd Maryland reported losses of 21 killed, 66 wounded, and 58 missing for a total of 145 casualties between August 16 and September 2, 1862.[12] The regiment was also on the field for the Battle of Chantilly on September 1, 1862, the largest engagement in Fairfax County, Virginia.

Thomas came through both these battles unscathed and was present for the regiment's next battle — at Antietam as described above. Probably sometime after the battle of Antietam, Thomas bought his disc since he had "Antietam" stamped on it. This can be seen in the photographs of his disc. (See page 112.) It is a sub-style (1A), one of several that bear the image of Major General George McClellan on the obverse. Thomas remained with the 2nd Maryland until his records revealed, "Noted on April 23, Absent with out leave 17 days." Private Tracy returned to the regiment and was reported as "present" from May to December, 1863. Like many of his comrades, Thomas mustered out on December 31, 1863, to allow his re-enlistment as a Veteran Volunteer and also so he could receive a $300 bonus and a thirty day furlough. Private Tracy was "present" during January and February 1864 before he was allowed to take his furlough and is shown as "absent" on furlough during March and April. After this Thomas returned to the regiment and was "present" during May and June, 1864 before he became sick and was sent home sometime in July or August as his records show him "absent" for this period. The last note in his file is that he died Sept 26, 1864 at home.[13]

The second soldier to enlist was Private James Kirby of Company I of the 6th New Hampshire Infantry. His identification disc, an Eagle (5A), is a common style that was sold to soldiers as early as February 1862.[14] Note in the photographs that it presents a problem that is occasionally found with these personal items. (See page 87.) Perhaps the regiment's sutler, W.M. Johnson, was in a rush because the sutler didn't get all the information correct.[15] The reverse of the disc identifies the owner as "JAMES JERBY CO. I 6. REG N.H.V. CONCORD." However, there was no soldier in the 6th New Hampshire named James Jerby. There is little doubt that the disc was period, so perhaps the owner's last name that was misspelled. So to whom did this ID disc belong?

Fortunately, the regiment's published history contains a roster. Assuming that only the last name was misspelled means that the search should be for someone in Company I from Concord with a first name of James as was stamped on his disc. Going through the roster of Company I and listing everyone with a first name of James produced the following sixteen candidates:

	Birthplace	Residence at time of enlistment
Sergeant James C. Dame	Dover	Concord
Private James Collins	Germany	Francestown
Private James Curley (alias James Ryan)	Mass	Hebron
Private James Dinwiddie	Harrison O.	Peterborough
Private James M. Hook	Concord	Concord
Private James T. Holbrook	Windsor N.S.	Alton
Private James Kirby	Concord	Concord
Private George W. Koller (alias James Williamson)	Cleveland O.	Stafford
Private James Lavell	England	Francestown
Private James W. Marden	Epsom	Epsom
Private James Manley	Gibson Ind.	Hill
Private James Murphy	England	Loudon
Private James C. Rand	Concord	Concord
Private James M. Sanborn	Sanbornton	Sanbornton
Private James Sullivan (alias James Scannell)	Ireland	Lisbon
Private James C. Whittle	Vermont	Keene

If the hometown is correct, the list is reduced to four names (shown in **bold**). Of these four one "James Kirby" seems most likely to be the correct soldier.

Three factors led to this name among the four. First, it is the only one with five letters in the last name — the others all have four letters. Second, the last three letters are the same as what was stamped on the disc, "rby." Finally, if the soldier spoke with an accent when saying his name, someone listening may have misunderstood or if it was written it could have been the handwriting "Je" vs "Ki." But regardless the owner of the disc was probably James Kirby.

Having concluded that the identification disc in question belonged to Private James Kirby, where and when did he serve his country? According to his service record Kirby enlisted on December 3, 1861, in Company I of the 6th Regiment New Hampshire Infantry for three years.[16] His age was given as 18. Private Kirby was present with his Company from January 1862 until February 1863 based on the bimonthly returns. Private Kirby was on the field at the battles of Second Bull Run, Chantilly, South Mountain and Antietam. Private Kirby was one of the lucky ones who survived them all in one piece and apparently without receiving any wounds. His records next show him as "Wounded slightly" on December 14, 1862, at Fredericksburg. On April 8, 1863, he was reported as "left in Hospital at Lexington Ky." He remained in a hospital until January 2, 1864, when he was mustered out to allow his re-enlistment as a "Veteran Volunteer." After James re-enlistment he returned to duty with the unit. He was present until May 12, 1864, when he was again wounded at Spotsylvania Court House. On June 7 he was described as "readmitted to hospital with Fever." He was absent in the hospital until November 1864, when he returned to duty. He saw further action at Petersburg and was present at Appomattox Court House. Private Kirby was mustered out with the unit on July 17, 1865, in Alexandria Virginia. Apparently, he escaped any long-term disabilities as a result of his service since he never filed for a pension. The Regimental History, which was published in 1891, stated (as of that date) that he was living in Cincinnati, Ohio.[17]

On October 1, 1861, twenty-five-year-old John W. Jenkins mustered into Company F of the 48th Pennsylvania Infantry as a First Sergeant[18] He very likely thought that soldiering couldn't be any harder or more dangerous than the mining which had been his previous job.

Sergeant Jenkins seems to have survived Second Bull Run, Chantilly and South Mountain without a scratch. On the morning of September 17, 1862, he and his comrades received orders to attack the small stone bridge over Antietam Creek which soon became known as "Burnside's Bridge." Fortunately for Sergeant Jenkins, his regiment was assigned to provide covering fire for the 2nd Maryland and 6th New Hampshire who were tasked with taking the bridge.[19] From a farm lane along side the creek, and later from a plowed field behind the rest of their brigade, the 48th attempted to suppress the deadly Confederate rifle fire poured down by the 2nd and 20th Georgia Infantry regiments positioned on a bluff high above the far side of the creek.[20]

The 48th was severely abused by the artillery of J.S. Brown's Virginia Battery, who reportedly included pieces of railroad iron as long as fifteen inches in their barrage.[21] Most likely at this point Sergeant Jenkins was struck in the head by a jagged piece of shrapnel that fractured his skull. At David's Island Hospital in New York City, A.A. Surgeon William K. Cleveland recorded that he "...applied the trephine to the point of fracture, and removed two pieces of the depressed internal table, an inch and a quarter in diameter which was followed by the escape of a few drops of pus." Sergeant Jenkins' war was over and he received a medical discharge on November 25, 1862.[22]

It may be assumed that while below the heights on the north side of Burnside's Bridge, and while recuperating in the hospital, Sergeant Jenkins wore his McClellan (1B) identification disc around his neck. Since his disc included a reference to his being an Orderly Sergeant, a promotion that didn't arrive until some time in May or June, 1862, it may also be assumed that he purchased his disc from W.M. Clarke, the 48th's sutler, shortly thereafter.[23] Judging from the moderate wear on Jenkins disc, which could hardly have been accumulated in just six months, the Sergeant probably wore or carried his disc long after discharge. Perhaps he was convinced that it brought him good luck.

The fourth and last of the four disc wearing enlistees to fight in front of Burnside's Bridge was Private Charles L. Perry of Company D of the 9th New Hampshire Infantry. How did he get to Antietam? The 9th New Hampshire mustered 900 young men into the service of the United States between July 3 and August 23, 1862.[24] The regiment arrived in Washington, D.C., on August 27, 1862, and was promptly assigned to the defenses of Washington. When the Union Army was reorganized after the Battle of Second Bull Run, the 9th New Hampshire was assigned to Colonel James Nagle's First Brigade of the 9th Corps on September 6, 1862.[25] As such, the regiment joined a veteran brigade. Private Perry's service records show that he mustered in on August 9, 1862, in response to President Lincoln's call for 300,000 additional troops in the summer of 1862.[26] Charles joined Company D at Concord, New Hampshire. He enlisted for three years and was paid a bounty of $25. Unfortunately, his service records failed to include any personal description of him at the time of his enlistment.

Perry was recorded as "present" from August 23, 1862, through February, 1863. At Antietam, the 9th was placed in support of the attack made by the 2nd Maryland and the 6th New Hampshire. The 9th eventually crossed the bridge following the successful attack by Brigadier General Edward Ferrero's Brigade. In the fighting, the regiment suffered 10 men killed and 49 wounded.[27] Private Perry was not one of the casualties so he survived his first two combat actions at South Mountain and Antietam in the first five weeks of his service.

Probably in the fall of 1862, Charles bought the brass identification disc, an Eagle (5A) substyle shown in the photographs. (See image on page 88.) He may have purchased it before the Battle of Antietam while the regiment was stationed in Washington, D.C. This sub-style of identification disc was being sold as early as February 1862 to other units so it is possible Charles bought it as soon as mustered into the regiment.[28] Note that it is the same style disc as Private Kirby's disc. New Hampshire soldiers commonly purchased this style of disc.

While the 9th was moving west, Private Perry decided he had had enough of the war. The bimonthly report for March/April 1863, revealed that he "deserted on the march from Boonsborough, Ky. April 17, 1863, since arrested and confined Richmond Ky." For May/June 1863, Private Perry was "held by provost guard awaiting sentence of court martial." Apparently he was returned to the regiment as he is shown as "present" but "under arrest" until November/December 1863, when he was reported as "sick in post hospital." The first hospital identified was at Paris, Kentucky. The records show that later in March or April 1863, he was at a hospital in Annapolis, Maryland, until July/August 1864, when he was returned to his regiment.

Meanwhile, the 9th Corps had returned to Virginia and was engaged in the fighting as the Army of the Potomac moved south toward Richmond in what is now known as the Overland Campaign. On September 30, 1864, the 9th New Hampshire was involved in the fighting at Poplar Spring Church. Brigadier General Simon Griffin, the commander of the brigade to which the 9th New Hampshire was attached at that time wrote in his report "I formed my command in two lines of battle, the Ninth New Hampshire on the right of the Eleventh Hew Hampshire.... I advanced steadily, driving the enemy's skirmishers before me.... Orders came to advance. I obeyed the order, but on arriving at the Jones House we met the enemy also advancing, with a line of battle stronger than our own, and overlapping us on both flanks."[29] Later in the report he wrote "...attacked on three sides by superior numbers they were [the brigade] compelled to abandon the place, [Jones House] losing heavily and having some of their men captured." The report shows the 9th reported losses of one killed, 20 wounded and 101 missing.

Among the missing was Private Perry. His service records continue to list him as "Missing a Prisoner of War" until the unit was mustered out on June 10, 1865, when it was noted that the government stilled owed him $75 of his bounty. It is not clear if Perry had returned by the time the unit mustered out, but probably not, based on the last item in his service records, a "Memorandum from Prisoner of War Records." It lists Private Perry as missing, but includes this very interesting notation, "...joined the Rebel army while a Pris. Of War, at Salisbury N.C. date not given." So far, no soldier has been identified in the Confederate Army by the name of Charles Perry in any unit stationed in Salisbury, North Carolina.

So Charles's story ends on a note of mystery. Is the notation true? Perry never filed for a pension and the regimental history contains nothing besides his being reported missing. So what happened to him? Did he really join the rebels? Did he die in a prisoner of war camp? Why didn't he return to New Hampshire after the war? We will probably never know.[30]

So we are left with four small pieces of metal that represent four soldiers each of whom met a different fate during the war but all of whom were present on September 17, 1862, at one of the most famous locations in the Civil War. Perhaps each of them wore his disc on that September day.

Chapter Notes

Chapter 1

1. Hankinson, pp. 33–86.
2. Jessup, *Vol. II*, p. 706. Imagine the shock if 18,000 (thirty percent of the 55,000 who eventually died in Vietnam) had been killed in the first battle in Afghanistan or Iraq.
3. No official identification discs, or dog tags as they are now called, were provided by the U.S. government until General Order No. 204, issued on December 20, 1906, mandated aluminum tags roughly the size of the discs purchased during the Civil War. Braddock, p. 15.
4. Haydon, *An Identification Disc for the Army, 1862*, pp. 62–63.
5. The insert pictured reads "*Bible House, Baltimore*. From the Maryland State Bible Society. To _____ Soldier in Comp'y. _Reg._____Vols. Should *I die on the* battle field or *in* Hospital, for the sake of humanity acquaint _____ residing *at* _____ of the fact *and where my* remains may be found." (Emphasis supplied from source other than figue shown.)
6. At the Battle of Cold Harbor in 1864, many Federal soldiers became nearly desperate in their desire for some way to identify their remains if the worst occurred. A young Union officer writing thirty years later recalled walking down the lines and seeing many of the men with their sack coats off, busily sewing. He wrote, "This exhibition of tailoring seemed rather peculiar at such a moment." On closer observation, he saw that "the men were calmly writing their names and home addresses on slips of paper and pinning them on the backs of their coats, so that their bodies might be recognized and their fate made known to their families at home." Porter, pp. 174–175.
7. In the spring of 1863, Major General Joseph Hooker assigned geometric symbols to each of his corps. For example, the First Corps was given a circle while the Sixth Corps a Greek Cross. After the soldiers began to identify with their symbol, merchants, and perhaps sutlers, began to offer corps badges in the various shapes. These badges were readily adapted for identification and commonly bore the owners' information engraved on the front. Very few corps badges had the identification information stamped in a manner similar to identification discs. (See the photograph on page 8.)
8. Faust, *Civil War Encyclopedia*, pp. 456, 662. The number "(12A)" refers to the authors' catalog number for that substyle disc provided in Chapter 2.
9. It has been alleged that some colonels granted permission in exchange for cash bribes, free liquor, or both.
10. Lord, *Civil War Sutlers and Their Wares*, pp. 24–25.
11. See Chapter 1, p. 9, and Chapter 2, p. 53.
12. Zeller, "Lying at the Fire Asleep," p. 5.
13. Stanley Phillips, *Excavated Artifacts*, p. 118. Kearney and McClellan suspension pins, with identification discs, were sold at the Stack's January 18, 2005, auction. See auction catalogue, pp. 223, 324. The disc was commonly joined to the suspension pin by a wire clip similar to a tightly closeable Christmas ball wire hanger or a fishing swivel. At least one Yankee had his picture taken with his disc attached to a button hole by a ribbon and another wore one on a his pocket-watch chain. (See the photograph on page 11.) The McClellan suspension pin was included in the Sullivan/DeWitt book *American Political Badges and Medalets 1789–1892* under their number GMcC 1864–26 with the added the note that the pins were manufactured in Boston (p. 313).
14. White metal will have to suffice as a general description since it is beyond the ability of the authors, and presumably most collectors, to distinguish between the various types of white metal used, which included silver, pewter, German silver, base metal, nickel, tin, and lead. It is believed, however, that the original Washington (3A) discs, and a majority of the nonbrass styles, were made from pewter.
15. One identification disc has been discovered where the soldier put a very bold X across McClellan's bust. (See Chapter 3.)
16. Doud and Jenkins Military Records.
17. Finch Military Records.

18. McInnes Military Records.
19. Mitchell Military Records.
20. Haydon, p. 63.
21. Ibid., p. 63.
22. Fox, *Regimental Losses*, p. 541.
23. Many were probably listed as missing in action at the Wilderness because their remains were burned beyond recognition in a fire that swept through the underbrush on the battle's second day. Presumably the change in the ratio between dead and missing was inverted at Spotsylvania because, until they moved south, the Federals held the field for several days.
24. Fox, *Regimental Losses*, p. 541.
25. Graham Military Records; American Civil War Research Database–Boshart. Parenthetically, the same obverse Union design was muled with an obverse die that bore the bust of Lincoln and the words "ABR'M LINCOLN a foe to traitors." It does not appear that this medal was intended for identification purposes since it lacked both a hole for suspension and a blank area for an inscription. It is possible this style was first used as a campaign medal in 1864, with the reference to Lincoln removed later so the design could be used as for identification discs.
26. Cooper, Bickel & McDevett Military Records.
27. Phillips, *Corps Badges*, pp. 145–146.
28. Goodwin, *Team of Rivals*, p. 418.
29. Phillips, *Corps Badges*, pp. 145–146.
30. *Harper's Weekly*, Vol. V, #241, p. 511. Presumably this advertisement was run by William A. Hayward of 208 Broadway, Providence, R.I. On April 2, 1864, Hayward also promoted his "Corps badges, staff badge for Sherman, new Grant, Hancock & Custer medal; gold & Silver badges." Bazelon and McGuinn, p. 119.
31. *Harper's Weekly*, Vol. V, #258, December 7, 1861, p. 784.
32. *Frank Leslie's Illustrated Newspaper*, Vol. XIII, #324, February 8, 1862, p. 191.
33. Courtesy of Lewis Leigh, Jr. Mr. Leigh has Henry T. Blanchard's disc as well as the original letter quoted. The authors have never seen one of these discs offered for sale. It is possible, although no evidence in support has been found, that other individual soldiers became agents who sold manufactured identification discs to their comrades rather than the regimental sutlers. It was not unheard of for enterprising soldiers to make spending money on the side by selling various items. One such product was a poster titled a "Soldier's Memorial," which was printed with patriotic motifs and information about a particular regiment and then sold in camp by soldier/agents. Another example of individual enterprise is found in *A Grand Terrible Dramma*, a compilation of published letters, where the author, Charles Wellington Reed, described how he drew sketches of his battery's fight at Gettysburg, had them reproduced at home, and sold them to his comrades as keepsakes.

34. Phillips, *Civil War Corps Badges*, pp. 145–146.
35. Lord, *Civil War Sutlers*, pp. 58–60.
36. Phillips, *Civil War Corps Badges*, p. 145.
37. Lord, *Civil War Sutlers*, pp. 59–60.
38. Rumors of the survival of original sutler punch kits surface from time to time. For example, one concerned an Iowa re-enactment regiment that had one and used it to make discs for the entire unit.
39. Occasionally these kits appear on the market and should not be confused with Civil War letter punch kits. As will be discussed in Chapter 4, the letter font style found on First World War letter punches was subtly different from those used in the Civil War as well as being approximately a third larger.
40. Lord, *Civil War Collectors Encyclopedia*, Vol. I, p. 132.
41. A review of disc samples from 1906 through World War I, where individual punches were used, shows similar but somewhat less precision as typically seen on Civil War discs, with some allowance for the fact that sergeants had more to do and didn't have to worry about paying customers. Braddock, pp. 15–36.
42. Ibid., p. 15.
43. Several Civil War relic dealers' websites have offered devices clearly intended to hold letter punches in a gang. It is doubted that these were intended for use with identification discs since the holder formed a straight line and because they were too large. Some similar but smaller device is conceivable.
44. No Civil War identification badges have been observed with the names and units stamped rather than engraved, and only one minted identification disc has been observed with engraving, and that only in a photograph on file at the United States Military History Institute in Carlisle, Pennsylvania.
45. Lord, *Collector's Encyclopedia*, Vol. II, p. 82. This story may have been embellished or may be wholly apocryphal. It is hard to imagine a colonel halting his entire regiment while his men bought identification discs. It must have taken several minutes each for the soldier to relay his information, the sutler to stamp the disc, and the soldier to pay. If any substantial number of discs were purchased, the regiment spent hours standing in line or resting beside the road while waiting for their turn. Perhaps the discs were purchased in the evening after the day's march was completed. Another version of this story has the regiment's colonel paying for the entire regiment's discs.
46. One of the authors has tried his hand with World War I punches with mixed success. Clearly, the sutlers invested significant time and patience in stamping discs, and probably only became relatively proficient after numerous attempts.
47. Yeomans, *1993 Handbook of United States Coins*, p. 5.

48. Al C. Overton, *Early Half Dollar Die Varieties 1794–1836*, 3rd ed., p.1. It is assumed that the gold and silver coating was applied by the manufacturer after the disc was minted.

49. Fuld, *Patriotic Civil War Tokens*, p. 188.

50. Smith's mark appeared in a least four combinations on McClellan style discs, "Smith F, Smith B, F.B. Smith, and B. Smith F." This variety indicates that Smith either had enough demand to require several dies, or his dies were weak and had to be replaced. On his Lincoln (6A) disc, Smith chose F.B. Smith.

51. David E. Schenkman, *The Numismatist,* pp. 788–816; Bazelon and McGuinn, *Directory of American Military Goods Dealers and Makers*, p. 186; Q. David Bowers, *Coin World*, p. 86.

52. Fuld, *Patriotic Civil War Tokens*, p. 184; Breen, *Breen's Encyclopedia of U.S. and Colonial Coins*, p. 666.

53. Bazelon, *Directory*, p. 169.

54. It is also possible that large manufacturers such as the Scovill Manufacturing Company commissioned dies from various die-sinkers and prohibited individual attribution.

55. The Displayed Eagle used for identification discs may have been traced from an actual coin, but subtle differences lead the authors to conclude the coins offered inspiration for free hand likenesses.

56. Bazelon, *Directory*, p. 239.

57. Rulau, *Tokens*, pp. 356, 370. Apparently Scovill considered making a knockoff of the United States Gold Eagle but based on legal advice, refrained from using both a liberty bust on one side and a displayed eagle on the other. Rulau, *Tokens*, p. 221.

58. Fuld, *Patriotic Civil War Tokens*, p. 186. As the Civil War progressed, coins became scarce and in order to make small change, merchants began to buy tokens that resembled one-cent coins as replacements. Some, known as store cards, had a patriotic symbol on the obverse and the merchants name, wares, and address on the reverse. Others, made strictly for use as substitute coinage, called patriotic tokes, usually held two patriotic images or phrases but no advertising.

Denticles are toothlike decorations extending from the rim into the field. Scovill's decoration resembled a string of pearls half pressed into the surface (see Chapter 2). Almost all Civil War identification discs included either denticles, beads or stylized ropes around the rim.

Denticle seems to be the accepted term for the small toothlike decorations and is the form used by the Fuld brothers, but at least one coin expert called them dentils. Breen, *Complete Encyclopedia*, p. 699.

In a telephone interview kindly granted by Bruce Bazelon, he related that at one time he was given charge of all of Scovill's paper records and remaining dies. He said that no significant records for the company dated before 1864 survived except for some correspondence, and that the dies were all for buttons. The paper records were donated to the Harvard Business School and the University of Connecticut while the button dies went to eight museums. He did not recall seeing any identification disc dies or any reference to same.

59. Due to the minters' general indifference to the quality of the strike, it is difficult to be 100 percent certain other bead borders were not intended. However, of the examples available to the authors, only the Shield (2B) may also have beads rather than denticles.

60. To date the authors have been unable to identify with any confidence other disc minters except, perhaps, the named die-sinkers themselves.

61. Fuld, *Patriotic Tokens*, pp. 63, 175–176.

62. Schenkman, *Civil War Sutler Tokens*, p. 76. Childs worked in Chicago, Illinois, and supplied mostly western theater Union regiments. Schenkman listed a number of items produced during the war by Childs, but identification discs were not among them. Sutler tokens were similar to patriotic tokens except the former usually bore the name of the sutler and were intended to secure the future business of those who got them in change.

63. Few patriotic or store card tokens had the die-sinkers name or initials included on the die either, leaving the Fuld brothers to make attributions based on laborious comparisons of letter styles.

64. Fuld, *Patriotic Tokens*, p. 50.

65. Ibid., pp. 50, 181.

66. Ibid., pp. 68, 178–179.

67. Use of identification discs was much less prevalent on the Western front, perhaps because of the relative poverty of soldiers from the country's newer states, and also because the Federals were able to bury their own dead with much greater frequency than their peers in the East. Discs continued to be sold up to the very end of the war but seeming with much less frequency than in the winters before the 1862, 1863 and 1864 campaigning seasons.

Chapter 2

1. The catalog of styles and substyles by basic description and number is arbitrary and was created to provide a more convenient and standardized designation for each particular style or substyle of disc.

2. Of course the authenticity of the engraving or stamping would still remain at issue.

3. It is likely that some of the dies used to mint identification discs were also used for medals, campaign badges (particularly the McClellan, Lincoln and Grant styles) and merchants tokens.

4. The Eagle (5D) is an exception, being a brass disc marketed with a silver wash rather than gold (numerical references are from the style catalog). One collection has been represented as containing a

Shield (2C) that was also coated with silver rather than gold, but this has not been verified because all specimens available for inspection by the authors were too heavily worn to determine the composition of the coating, if any.

5. No known and verifiable discs were made from aluminum during the Civil War because at that time the metal was still prohibitively rare and expensive. Aluminum did not come into common use until the late nineteenth century.

6. It is believed that virtually all "white metal" discs (except those made entirely from silver) are probably pewter and were originally coated with a silver wash much like the brass ones were coated with gold.

7. The authors have observed crossover identification discs. Examples include: McClellan (1A) in white metal; (1B) in copper; Shield (2A) in both copper and white metal; Shield (2C) in both white metal and copper-nickel; Eagle (5A) in both silver and copper; Lincoln (6A) in copper and white metal; and the Grant (9A) in both silver and white metal and the (9B) in both brass and white metal (all numeric references are from the style catalog). All crossover metal discs should be considered extremely rare or perhaps unique.

8. For example, there may be a hoard discovered somewhere that consists solely of one substyle and which were designated here as "extremely rare" only because the collector managed to corner the market.

9. Perversely, discs offered online also present the greatest concern about authenticity since the authors did not have the opportunity conduct a hands-on inspection.

10. It would have been preferred to use a rarity scale similar to that developed by the Fulds for patriotic tokens and store cards, where they projected and classified each token style's rarity as a total number existing in that universe, i.e. one to five examples known to exist. Unfortunately, neither sufficient perseverance nor sufficient numbers of identification discs were available to make such projections.

11. Lest the reader be misled by the term "common," it must be remembered that it is only applicable in comparison to other identification discs. At any given Civil War relic show one is likely to find hundreds of pistols, rifled muskets, swords, images and buttons, but only a half-dozen identification discs of all substyles, even on a good day.

12. Most identification discs have raised rims and a series of tiny, toothlike decorations around the inside of the raised rim called denticles, on both the obverse and reverse. Hereafter specific mention will be made of only those that either lack denticles or have some other type decoration. Also, for convenience, the reference to the die-sinker's identification mark will be abbreviated as "ds." Finally, some identification discs contained in this style catalog also appear in the text *American Political Badges and Medalets 1789–1892* by J. Doyle Dewitt as revised by Edmund B. Sullivan. Sullivan explained that the dies for certain identification discs were also used in political campaigns and vice versa (p. 271).

13. All of the McClellan (1B) discs personally observed by the authors displayed on the obverse two fused denticles above the R in "Major" and what appears to be a die crack from those fused denticles to the top of the R. There may be (1B) discs made from a different die without those markings and with a star between the words "War" and "Major" that is lower in position than on the die that has the fused denticles. This other possible die has a die-sinker mark as "F.B. Smith" rather than the B. Smith F. on the personally observed discs. Finally, in every instance where more than one metal has been listed, the first is the predominant metal found.

14. In *American Political Badges* Sullivan includes the following in reference to his #GMcC 1864-2 which bust bears a striking resemblance to that appearing on the McClellan (1E, F,G, H and M) discs: "Note: The dies were made by George H. Lovett of New York. Issued originally as a war medal, it received distribution during the presidential campaign. Practically all pieces of this size and smaller which were originally intended as war medals were distributed as political pieces, particularly among the soldiers of the Army of the Potomac" (p. 306).

15. All listings marked with an * are only tentative pending confirmation, since no stamped or engraved examples have been found.

16. A question mark appearing in place of weight, size or metal composition within the parentheses indicates that information for this substyle was gathered from a text or from an old, respected collection but where the authors have not personally had the opportunity to physically measure or weigh the disc or such information was not available from accepted sources.

17. DeWitt/Sullivan attributed this style and McClellan (1J, 1K and 1L) to die-sinker George Lovett in 1864. They also agree that these were sold as identification discs (DeWitt/Sullivan, p. 308).

18. "Contemporary reports indicate this was a popular piece with soldiers of the Army of the Potomac who were furloughed or invalided home" (DeWitt/Sullivan, p. 309).

19. Rulau/Fuld, *Medallic Portraits of Washington, 2d ed.* The authors of that text provided that the "War of 1861" usually found on the reverse of the Washington (3A) discs was included during minting on the 122T and omitted on the 122U (p. 109). These authors believe the phrase was stamped on by sutlers based upon minor variations in location on the reverse, the occasionally observed splitting of the words "war of" and the date, and the halo effect common to the other added information.

20. Although both this disc and the Washington (3A) have the same rating for rarity, it should be noted that almost 50 percent more 3As appear in the survey list in Chapter 5 than do Washington (3C)s.

21. See Chapter 4, page 155. Any 3D style disc that has a reverse with words and lines minted onto the surface is a restrike done for the First World War using Lovett's original obverse die. Any reference to "Baker 621" being from the Civil War applies to only a disc with a blank reverse without words and lines. Certain texts may be in error on this point.

22. See Chapter 4, page 153. Well struck minted reproductions exist, usually with sections of the arrow shafts missing and a tiny word "copy" on the reverse.

23. The bust of Lincoln is similar to DeWitt/Sullivan #AL 1864-3 and is also attributed to die-sinker Frederic B. Smith, but those authors do not include a disc identical to 6A in their text.

24. This obverse is the same as the reverse of DeWitt/Sullivan #AL 1864-8.

25. The only specimen observed has engraved biographical information rather than stamped. This may be attributable to the fact that these were apparently sold by advertising to jewelers rather than to sutlers

26. DeWitt/Sullivan provided the following regarding a possible fourth Grant style identification disc under #USG 1868-9: "Probably this piece was struck originally as a 'War medal' as pieces with the same incomplete reverse are found with an obverse of General William Sherman and General Winfield Hancock" (p. 344). Neither of the authors of this book has observed such a disc either stamped or in blank.

27. This obverse was muled (combined) with a reverse reading "Joint Committee on Escort of Corn Exchange and Citizens for 2nd Regt. Blue Reserves." It is impossible to determine at this time whether the obverse was designed for an identification disc first and then appropriated as a medal or vice versa.

28. DeWitt/Sullivan attributed this obverse to die-sinker John D. Lovett and added that it was made for an 1860 Lincoln campaign medal (p.263).

Chapter 3

1. Pennsylvania Reserve Volunteer Corps.

Chapter 4

1. NGC, P.O. Box 4776, Sarasota, FL 34230; (800) 642–2646; www.ngccoin.com/. An NGC spokesman has indicated that the output of certified discs has been fairly modest to date.

2. Civil War records were generally very good. However, mistakes were made, particularly in spelling. Also, men often changed companies, and sometimes regiments, after purchasing a disc. When attempting to evaluate disc information that doesn't match the record, it is suggested that the searcher check all possible variations in spelling and all other companies for similar names. There are even some men whose names were lost for various reasons. One example of the latter is found in the records of the 12th Pennsylvania Volunteer Cavalry, where many of the regimental muster records were lost in June of 1863 at the Second Battle of Winchester. There were a number of men killed and captured from the 12th whose records were lost and never recreated because they never rejoined the regiment.

3. Currently the contact address is "Old Military and Civil Records (NWCTB-Military), Textual Archives Services Division, National Archives and Records Administration, 700 Pennsylvania Avenue NW., Washington, DC 20408-0001." The National Archives may also be reached on-line.

4. National Park Service Civil War Soldiers and Sailors System, http://www.itd.nps.gov/cwss/index.html. Many states provide on-line access to their archived information on Civil War soldiers, also for free.

5. American Civil War Research Data Base, www.civilwardata.com. At the time of writing, the annual fee was $25.00 per year as of early 2008.

6. Photographs of referenced discs can be observed in Chapter 3.

7. Edward must have acquired his disc after that date.

8. The belief that some battles came as punches with whole words is based primarily upon the appearance of the letter alignment along the bottom of the words, which were much more uniform than would have been expected from a sutler using tiny individual letter punches.

9. The determination that the listed battles were stamped with gang letter punches is supposition based upon the uniformity of the spacing, alignment of the letters, and uniformity from disc to disc. However, the existence of gang letter punches should not be taken as definitive since no such punches have been observed by either author.

10. Many old coins have minor variations in small details such as placement of stars, appearance of numbers, etc., because the technology of the time resulted in significant die breakage and replacement. Some coins had years with as many as a dozen different dies. Because of the methodology, there was usually some variation from die to die. The authors do not purport to have observed enough discs to determine the full extent of such minor die variations as have been identified for old coins and for Civil War patriotic and merchants tokens (although die variations will be noted when observed).

11. One of the keys to detection of fake discs is that some are obviously cast and all have far more wear than would be the case if engraved during the

Civil War, when the coins would have been almost new. Presumably to save money, the fakers bought and used heavily worn coins for use or casting because they were cheaper.

12. Hartzog, Rich, AAA Historical Americana-World Exonumia, Exonumia.com/Fakes.htm. Hartzog's web site provides the following observed variations of the bogus "Pender" discs along with excellent photographs: "Lt. Col Dorsey Pender — CSA — 1st North Carolina" on a counterfeit 1860-Oo U.S. half dollar; "Lt James B. Washington" on an 1862 half dollar; "A.L.P. Vairin — Second Mississippi" on an 1861 half dollar; "Robert S. Bunting — CSA ([on eagle)] — 8th Texas Cavalry" on an 1862-Ss half dollar; "Col. Joseph Mayo — 3rd Va. Inf." on a counterfeit 1860-Oo half dollar. Hartzog also provided other observed names for which he apparently lacked pictures; Thomas Taylor, William Forrest, and Nathan Boone of the 12th Kentucky.

The "Mayo" disc in the illustration is at least predominantly silver, is cast, has no reeding around the outside rim, is 37 mm in diameter as opposed to the correct 38.5 mm, and weighs 19 grams. instead of the correct 26 grams (Breen, *Walter Breen's Complete Encyclopedia...*, p. 437). The opinion that the "Mayo" disc contains predominantly silver is based solely upon a ring test, not a chemical analysis.

13. The earliest example observed to date, a shield shape with an eagle motif, included on the reverse that it was patented on December 29, 1868. Apparently similar designs were also used as dog tags in the Spanish-American War.

14. Russell Rulau, *United States Tokens: 1700–1900*, p. 576.

15. The authors have observed at least one disc sold at an on-line auction several years ago where just such a ruse was almost certainly perpetrated. Sadly, some unsuspecting collector paid more than $1,000.00 for the item.

16. Perhaps the brass used in the knock-off is softer than the original since the reproduction has the same diameter and weight as the original. The authors have not observed authentic discs with similar markings.

17. At least one poorly cast copy of a Shield (2A) disc has been observed for sale on eBay, where World War I punches were almost certainly used on the reverse.

18. Thomas DeLorey article, *The Numismatist*, July 1980, p. 1630. The Elder restrike appears as Baker 621a in Rulau and Fuld's book, *Medallic Portraits of Washington...*, 2nd ed., on page 249 but is erroneously listed along with the Washington (3C) and (3D) under the broad heading of "Civil War Dog Tags." After personal communication, Mr. Rulau graciously acknowledged that including the Elder re-strike as Civil War identification discs was an error, and that the pictured example was probably just such a reproduction. The original listing in their book did, however, include the caveat "Robert Lovett and some successors made wide use of this obverse die down to 50 years after its appearance" (p. 249).

19. The reproduction cast aluminum McClellan (1J) is distinguishable from the original in that it weighs only 4 grams compared to 10 grams, and that the reproduction is slightly smaller at 33 mm.

20. A "strike-over" is a blemish caused when a bit of foreign material like lint, a metal chip or dirt gets between the die and the blank leaving an impression.

21. "Field" is the numismatic term for the blank areas between the rim, the lettering, and the device, while "device" is the corresponding term for the primary image on the coin or token.

22. A die-sinker's mark can often provide an excellent diagnostic clue for uncovering a cast reproduction. The die-sinker's mark was usually well protected by the device. Unless the whole disc is heavily worn, the mark will usually be crisp and easily readable. Because of their size, however, the mark on a cast reproduction will usually be illegible or consist of little more than an amorphous lump of metal.

23. It is believed that some soldiers wore their discs on watch chains or carried them in pockets as watch fobs, mementos or good luck charms long after the war.

24. One clear exception to this rule of thumb applies to discs taken out of use shortly after purchase, such as just after a soldier's death or discharge.

25. PCGS, *The Fundamentals of Counterfeit Detection*, www.pcgs.com.

26. The designation of white metal here covers a number of different substances and alloys but aluminum was not one of them. Because of its scarcity and the cost of refining at the time, aluminum was too expensive for use in identification discs. Care must be exercised because there are a number of modern reproductions made from aluminum that are difficult to distinguish from the actual white metal used, especially the McClellan (1I, 1J and 1K). With those, weight is the best authenticator since all aluminum reproductions encountered to date weigh only about half of the original. It is believed (without scientific basis) that the McClellan (1E, F, G, H); Washington (3A); Battles (4B & C); and Lady Liberty (15A) are probably pewter, while the rest are probably tin or some alloy of both.

27. Please reference Chapter 2 for several substyles of brass discs that were coated with silver rather than gold. No discs made from white metal and coated with gold have been encountered.

28. There are several non-destructive tests to determine the type of metal from which a disc is made, which in turn helps in authentication. Pewter at the time contained lead. With the dealer's permission, draw the rim of the disc across a piece of paper with

moderate pressure. Lead will leave a very distinct mark like a lead pencil. In order to tell toned copper, which was rarely used for identification discs, from brass, which was used predominantly, rub the edge with a fine grade sharpening stone. Brass will show up on the stone as yellow while copper will be red-pink in color. Cunningham, Paul, "Watch Out for This Scam," *The Civil War Token Journal*, Spring 2006, Vol. 40, #1, p. 4. Always ask the dealer's permission before subjecting his disc to either test. If one is checking the authenticity of a previously collected disc, the test spot will retone over time,. If there is not sufficient time for retoning, a fresh test mark should provide reassurance to a buyer because the exposed metal should have the appropriate color (imagine a new cent versus a brass door knob). If the disc is bogus, the damage won't matter anyway.

29. It should be noted that silver and silver wash will tone (i.e., tarnish) over time, changing from red, then to blue, and finally darkening to almost black. Dark toning on non-worn discs looks different than the toning on a white metal disc that has lost much of its silver coating. The former will have some shine beneath the toning whereas the worn disc will display a much flatter charcoal gray color.

30. It seems that silver held up better than gold since heavily worn examples of white metal discs seem to retain more silver wash than similarly worn discs with gold. Perhaps this occurred either because silver is harder than gold, or because it was less expensive and a thicker coating may have been applied.

31. This phenomenon is especially noticeable in the hair of a portrait. It is common to see shine in the grooves and part in the hair, but none on the ridges.

32. At least one experienced relic hunter has reported that new green corrosion can be distinguished by rubbing the corrosion. If it comes off with light pressure, it was probably artificially induced since the corrosion lacked the necessary time to deeply bond with the metal. If the corrosion doesn't rub off easily it is more likely genuine for the opposite reason.

33. In general it is not a good idea to dip a disc in any sort of tarnish remover to eliminate toning or stains because a pristine surface raises the possibility of a modern reproduction. Although surface condition is not as important as it is with coins and tokens, only excavated discs should be cleaned, and then only enough to remove the dirt.

34. Lack of a suspension hole is not necessarily an indication of fraud or of a reproduction. A few styles were apparently minted without holes, presumably to be carried in the pocket.

35. As may have been noted earlier, some of the identification discs came with hanger tabs, which show wear consistent with the rest of the disc. However, special attention should be paid to the area between the rim and the hole on hanger tabs on the Battles (4C) style because the words "Merriam" and "Boston" were minted into authentic identification discs. The markings may not be present or visible on heavily worn pieces but on lightly worn pieces the opposite should be true.

36. No such instructions have been found from the Civil War, but each World War One kit included a small diagram printed in the case in order to insure uniformity.

37. Another characteristic peculiar to Eagle (5A) style discs is that the decoration around the rim of the obverse consists of a string of beads. Virtually all other discs have denticles, ropes, rims or nothing. No Eagle (5A) modern reproduction has been observed with clearly delineated beads.

38. Oddly, it seems that kits usually came with separate punches to add behind the regimental number such as "TH," "ST" and "ND," which are almost always devoid of serifs. Those punches often included a line or double dots underneath. It is assumed that most of the punches held both the abbreviation and the line or dots, but there are some discs that appear as if the dots were just two periods.

39. The die-sinker mark on the Washington (3A) disc is not as well protected and its condition will not provide as strong an indicator unless entirely missing. Further, it appears that on some (3A) discs the mark was intentionally defaced on the die by being gouged away. A defaced mark is not necessarily a caution sign. Similarly, on the Battles (4C) the interior of the hanger tab often collapsed from heavy wear and obliterated the die-sinker mark.

40. It is always possible that the dealer bought the disc that day from a vest pocket dealer or attendee who didn't have the records, and may not have known what they had for sale. In that situation, probe the dealer and if any doubts remain, ask about a possible return if the name on the disc doesn't match or the soldier's record does not support the price paid.

41. The issue of price can be controversial. Some collectors will avoid all significantly suspect discs regardless of the price. Others may purchase a disc with some problems if a reasonable discount is offered to compensate for the additional reasonable risk being assumed. In general, it is probably safest for the beginning disc collector to pass on a questionable disc unless very confident of the dealer or the person providing a second opinion, and for everyone to avoid extremely cheap discs unless it is understood that the payment is the functional equivalent of dropping ones money into a slot machine.

42. In another scam the second high bidder is contacted and told that he can buy the item because the high bidder backed out. The simplest safeguard is to e-mail the listed seller for confirmation.

43. This statement should not be taken to imply such problems with other type relics—only that the authors lack the expertise to make judgments about other items.

Chapter 5

1. The fact that a particular identification disc is different from others from the same regiment or company does not necessarily mean that the disc is bogus. The disc could have been given to the soldier as a gift or the sutler may have changed style. Such a discrepancy should add a note of caution, however.

Epilogue

1. United States War Department, *The War of the Rebellion: A Compilation of the Official Records of the Union and Confederate*, Series 1, Vol. 19 , Pt. 1, p. 178.
2. *The Union Army*, Vol. II, p. 271.
3. *The Union Army*, Ibid., Vol. I, p. 84.
4. Ibid., Vol. I, p. 380.
5. Ibid., Vol. I, p. 84.
6. United States War Department, *The War of the Rebellion: A Compilation of the Official Records of the Union and Confederate*, Series 1, Vol. 19, Pt. 1, pp. 446–447.
7. United States War Department, *The War of the Rebellion: A Compilation of the Official Records of the Union and Confederate*, Ibid., Series 1, Vol. 21, p. 132.
8. *The Union Army*, Vol. I, p. 85.
9. Ibid., Vol. I, p. 380.
10. Tracy, Thomas military records.
11. Hennessy, *Return to Bull Run: The Campaign and Battle of Second Manassas*, p. 263.
12. United States War Department, *The War of the Rebellion: A Compilation of the Official Records of the Union and Confederate Armies*, Series 1, Vol. 12, Pt. 2, p. 261.
13. Tracy, Thomas military records.
14. Zeller, "Lying at the Fire Asleep," p. 32.
15. Lord, *Civil War Sutlers...*, p. 101.
16. Kirby, James military records.
17. Jackson, *History of the Sixth New Hampshire Regiment in the War for the Union*, p. 410.
18. Jenkins, John W. military records.
19. Sears, *Landscape Turned Red*, p. 264.
20. Priest, *Antietam...*, pp. 225–310.
21. Ibid.
22. Jenkins, John W. military records.
23. Lord, *Civil War Sutlers...*, p. 108.
24. *The Union Army*, Vol. I., p. 85.
25. Lord, *History of the Ninth Regiment New Hampshire Volunteers in the War of the Rebellion*, p. 41.
26. Perry, Charles L. military records.
27. United States War Department, *The War of the Rebellion: A Compilation of the Official Records of the Union and Confederate*, Series 1, Vol 19, Pt. 1, p. 197.
28. Zeller, "Lying at the Fire Asleep," p. 32.
29. United States War Department, *The War of the Rebellion: A Compilation of the Official Records of the Union and Confederate Armies*, Series 1, Vol., 42, Pt. 1, pp. 587–588.
30. Perry, Charles L. military records.

BIBLIOGRAPHY

Manuscript

Blanchard, Henry T. Letter, July 29, 1862. Courtesy of Lewis Leigh, Jr., Collection.

Books, Primary

Jackson, Capt. Lyman. *History of the Sixth New Hampshire Regiment in the War for the Union.* Reprint, Salem, MA: Higgisons Book Company, 1891. First published by Republican Press Association.

Lord, Edward O. *History of the Ninth Regiment New Hampshire Volunteers in the War of the Rebellion.* Reprint, Salem, MA: Higgisons Book Company, 1895. First published by Republican Press Association.

The Union Army. Wilmington, N.C.: Broadfoot, 1997. First published 1908 by Federal Publishing Company.

United States War Department. *The War of the Rebellion: A Compilation of the Official Records of the Union and Confederate Armies.* Washington: Government Printing Office, 1880–1901.

Books, Secondary

Bazelon, Bruce S., and William F. McGuinn. *Directory of American Military Goods Dealers & Makers, 1785–1915, Combined Edition.* Manassas, VA: Typesetting & Publishing, 1999.

Braddock, Paul F. *Dog Tags: A History of the American Military Identification Tag, 1861 to 2002.* Chicora, PA: Mechling, 2003.

Breen, Walter. *Walter Breen's Complete Encyclopedia of U.S. and Colonial Coins.* New York: Doubleday, 1988.

Campbell, Eric A., ed. *A Grand Terrible Dramma: From Gettysburg to Petersburg, the Civil War Letters of Charles Wellington Reed.* New York: Fordham University Press, 2000.

Crouch, Howard R. *Civil War Artifacts: A Guide for the Historian.* Fairfax: SCS Publications, 1995.

Faust, Patricia L. *Historical Times Illustrated Encyclopedia of the Civil War.* New York: Harper Perennial, 1991.

Foote, Shelby. *The Civil War: A Narrative — Fort Sumter to Perryville.* New York: Vintage, 1986.

Fox, Lt. Col. William F. *Regimental Losses: Regimental Losses in the American Civil War 1861–1865.* Albany, NY: Albany Publishing Company, 1889.

Fuld, George, and Melvin Fuld. *Patriotic Civil War Tokens.* Iola, WI: Krause, 1991.

Furgurson, Ernest B. *Not War but Murder: Cold Harbor 1864.* New York: Knopf, 2000.

Goodwin, Doris Kearns. *Team of Rivals: The Political Genius of Abraham Lincoln.* New York: Simon & Schuster, 2005.

Hankinson, Alan. *First Manassas 1861: The Battle of Bull Run.* Oxford, UK: Osprey, 2000.

Hennessy, John J. *Return to Bull Run: The Campaign and Battle of Second Manassas.* New York: Simon & Schuster, 1993.

Jessup, John E., ed. *Encyclopedia of the American Military*, Vol. II. New York: Charles Scribner & Sons, 1984.

Lord, Francis A. *Civil War Collectors Encyclopedia, Vols. I–V.* Edison, NJ: Blue & Gray, 1995.

_____. *Civil War Sutlers and Their Wares.* New York: Thomas Yoseloff, 1969.

Maier, Larry B. *Rough & Regular: A History of Philadelphia's 119th Regiment of Pennsylvania Volunteer Infantry, The Gray Reserves.* Shippensburg, PA: Burd Street Press, 1997.

Overton, Al C. *Early Half Dollar Die Varieties: 1794–1836, 3rd ed.* Ann Arbor, MI: Edwards Brothers, 1990.

Phillips, Stanley S. *Civil War Corps Badges and Other Related Awards, Badges, Medals of the Period.* Marceline, MO: Walsworth, 1982.

_____. *Excavated Artifacts from Battlefields and Campsites of the Civil War, 1861–1865.* Marceline, MO: Walsworth, 1986.

Porter, Horace. *Campaigning with Grant.* New York: Bonanza, 1961.

Priest, Michael John. *Antietam: The Soldiers' Battle.* New York: Oxford University Press, 1989.

Rulau, Russell. *United States Tokens 1700–1900*, 4th ed. Iola, WI: Krause, 2004.

_____ and George Fuld. *Medallic Portraits of Washington*, 2nd ed. Iola, WI: Krause, 1999.

Sauers, Richard A., and Capitol Preservation Committee. *Advance the Colors: Pennsylvania Civil War Battle Flags, Vols. I & II*. Lebanon, PA: Sowers, 1991.

Schenkman, David E. *Civil War Sutler Tokens and Cardboard Script*. Bryans Road, MD: Jade House, 1983.

Sears, Stephen W. *Landscape Turned Red: The Battle of Antietam*. New Haven, CT: Ticknor & Fields, 1983.

Sullivan, Edmund B., from J. Doyle DeWitt. *American Political Badges and Medalets, 1789–1892*. Lawrence, MA: Quartermans, 1981.

Woodhead, Henry, ed. *Echoes of Glory: Arms and Equipment of the Union*. Alexandria, VA: Time-Life, 1991.

Yeomans, R.S. *1993 Handbook of United States Coins, 15th ed.* Racine, WI: Western, 1992.

Articles

Bowers, Q. David. "Did Minter of Civil War Tokens Use Modular Dies for Multiple Values? *Coin World* (August 28, 2006).

Cunningham, Paul. "Watch Out for This Scam." *The Civil War Token Journal* (Spring 2006): vol. 40, no. 1.

DeLorey, Thomas. "Thomas L. Elder: A Catalogue of His Tokens and Medals." *The Numismatist* (July 1980).

Gross, John. "The Identification Disc of William D. Tanner and the Unheralded 21st New York Cavalry." *North South Trader's Civil War* (2002): vol. 28, no. 6.

Homren, Wayne, ed. "How Museums Handle Dies: The Scovill Die Experience." *The Numismatic Bibliomania Society: E-Sylum* (February 26, 2006): vol. 9, no. 9.

Lord, Francis. "Federal Army Identification Discs of the Civil War." *Military Collector & Historian* (March 1952).

Mullins, Tim. "An Artilleryman's ID Disc: John Burbeck, N.H. Light Artillery." *North South Trader's Civil War* (2001): vol. 28, no. 1.

Rossbacher, Nancy Dearing. "Identification Discs & Inscribed Corps Badges." *North South Trader's Civil War*. (1990): vol. XVII, no. 5.

Schenkman, David E. "Joseph H. Merriam — Die Sinker." *The Numismatist*. (April 1980): vol. 93, no. 4.

Scott, Will. "Deserters' ID Discs." *North South Trader's Civil War*. (May-June 1994).

Spangler, Norbert F. "ID Disc of James Denworth." *North South Trader's Civil War*. (2003): vol. 29, no. 6.

Stansbury, Haydon F. "An Identification Disc for the Army, 1862." *Journal of the American Military Institute*. (Spring 1939): vol. III.

Sylvia, Stephen W. "History Written Again at Stoneman's Switch." *North South Trader's Civil War*. (2006): vol. 31, no. 6.

Zeller, Paul G. "Lying at the Fire Asleep." *Military Images*. (July/August, 2002).

Newspapers

Frank Leslie's Illustrated Newspaper. February 8, 1862 (vol. XIII, no. 324).

Harper's Weekly: A Journal of Civilization. May 25, 1861 (vol. V, no. 230), August 10, 1861 (vol. V, no. 241), December 7, 1861 (vol. V, no. 258).

Military Records

United States National Archives:
Bickel, George.
Boshart, Jacob.
Cooper, Benjamin.
Dowd, Mortimer.
Finch, Lorenzo.
Graham, William.
Jenkins, John.
Kirby, James.
McInnes, James.
Mitchell, John.
Perry, Charles L.
Tracy, Thomas.

Index

Adams, Ara 88
Adams, Sylvanus 89
Adams Express 18
Albany, New York 98
Alexander, David D.P. 118
Alexandria, New Hampshire 89
Alexandria, Virginia 194
Allen, Joseph W. 129
Allendorf, George 128
Allentown, Pennsylvania 121
Aluminum 200, 202
Ambulance Corps 101, 111
American Civil War Research Data Base 201
Ames, Alfred 84
Andes, New York 120
Annapolis, Maryland 195
Antietam, Maryland, Battle of 13, 15, 25, 189, 190, 193, 194, 195
Antietam Creek, Maryland 189, 190, 194
Appomattox, Virginia, Battle of 192
Appomattox Court House, Virginia 194
Armstrong County, Pennsylvania 118
Army of the Potomac 6, 12, 189, 195, 200
Arrows 154
Augur U.S.A. General Hospital Alexandria, Virginia 84
Aumack, Thomas B. 107
Authentication: Casting Pits 156; Die-sinker Mark 164; Eagle (5A) 164; File Marks 157; Gold Gilding 164; Halo 164; Lettering 164; McClellan (1A) 164; Quick Check List 163; Shield (2A) 164; Silver Wash 164; Washington (3A) 164

B. & H.D. Howard 16
Baltimore, Maryland 189
Base metal 25, 197
Beach, George F. 94
Beauregard, Maj. Gen. Pierre Gustave Toutant 5
Belding, I.C. 18
Bellaire, Ohio 114
Belle Plain, Virginia 78
Berg, Frederick W. 138
Bibles 7
Bickel, George 15, 78
Bingham, Private 8
Bishop, Samuel R. 80
Bitner, Matthias 139
Blackhorse, Pennsylvania 77
Blanchard, Henry T. 17, 18, 198
Bleaderheiser, Andrew J. 133
Blemishes 157
Bone 7
Boone, Nathan 202
Boonsborough, Kentucky 88, 195
Boshart, Jacob 15
Bossart, David 123
Boston, Massachusetts 20, 82
Bowmansville, New York 84
Bradford, Vermont 146
Bradley, Julian A. (H) 117
Brass 25, 203
Brass stencil 7
Bremen, Germany 148
Bridgens, William H. 23
Brooklyn, New York 107
Brown, J. 91
Buffalo, Kansas 134
Bull Run Creek, Virginia 5
Bunting, Robert S. 202
Burlington, New Jersey 101, 102
Burnside, Maj. Gen. Ambrose 180
Burnside's Bridge, Maryland, Battle at 189, 190, 194, 195

California units: 1st Infantry 131
Camp Dennison, Ohio 113
Camp Randle Madison, Wisconsin 146
Canaan, Pennsylvania 134
Canton, New York 145
Cassady, John B. 133
Cast identification discs 157
Centerville, Virginia 5
Centre County, Pennsylvania 135
Chaffins Farm, Virginia, Battle of 77
Chancellorsville, Virginia 15
Chantilly, Virginia, Battle of 151, 189, 193, 194
Charlestown, West Virginia 20
Chase, George L. 85
Chicago, Illinois 23, 78, 199
Chickamauga, Georgia, Battle of 15
Childs, Shubael Davis 23
Cincinnati, Ohio 114, 194
Civil War Re-enactments 150
Clark, W.M. (sutler) 194
Cleveland, Surgeon William K. 194
Cliffbourne Hospital, Washington, D.C. 114
Clinton, Massachusetts 86
Clum, Chauncey 97
Coffin, James H. 119
Coins 7, 149
Cold Harbor, Virginia, Battle of 15, 192, 197
Colgan, Daniel 131
Color 159
Color and wear 158
Coloration 158
Commercial certification 149
Concord, New Hampshire 88, 193, 195
Connecticut 22, 25

207

Connecticut units: 8th Infantry 76
Conway, New Hampshire 88
Cooper, Benjamin 15
Cooper, William E 132
Cooperstown, New York 100
Copper 25
Copper discs 154
"Copy" 154
Cork, Ireland 105
Corrosion, soil 159
Cotton, Samuel H. 105
Counterfeit dies 158
Cullinan, John W. 82
Cumberland County, New Jersey 92
Cummings, Daniel 10
Cunningham, Paul 203
Custer Badge 19
Cuyler U.S.A. General Hospital, Germantown, Pennsylvania 115

Dansville, New York 105
David's Island Hospital, New York City, New York 194
Dealers 161
Delaware units: 2nd Infantry 77
DeLorey, Thomas 202
Denticles 24, 199
Detroit, Michigan 83
Device 202
Diarrhea 13
Dickey, John 79
Die-cutter 20
Die-sinker 20
Die-sinker marks 161; "B. Smith, F.B. Smith, F. Smith B., Smith F" 26, 27, 58, 199, 200, 201; Battles (4C) 161; "Boston" 51, 203; "G(eorge). H. Lovett" 48, 200; Lincoln (6A) 161, 199; "Lovett," and "R.L." 47; McClellan (1A) 161; McClellan (1B) 161; "Merriam" 161, 203; Washington (3A) 161
Dilts, David 92
Dog tag 9
Donath, Henry 82, 150
Doud, Mortimer H. 13, 137
Doughboys 19
Dunlap, Milton H. 121

Eckington Hospital, Washington D.C. 105
Elder, Thomas 155
Elder re-strike 155

Elkhart, Indiana 78
Elmendorf, Silas E. 134
Elmer, James P. 145
Elmira, New York 97, 98, 99, 110
Errickson, Elmer 92
Essex, New York 94

F.C. Key and Sons 23
Fairfax County, Virginia 193
Fairfax Seminary Hospital, Virginia 105, 116
Fellows, Col. E.Q. 190
Ferguson, Daniel 83
Ferrero's Brigade, Brig. Gen. Edward 195
Fields 157, 202
Finch, Lorenzo D. 13, 104
Finly Hospital, Washington D.C. 103, 111
1st Battalion Invalid Corps 115
First Division General Hospital Alexandria, Virginia 93
First World War identification discs 154, 201
First World War letter punch kit 20, 149, 154, 198, 202, 203
Flick, Peter H. 12, 104
Forbes, Edwin 10
Forrest, William 202
Fort Donelson, Tennessee 16
Fort Marcy, Virginia 113
Fort Monroe Hospital, Virginia 90
Fort Sumter, South Carolina 7, 9
Foss, Joseph H. 86
Fowler, A. 12
Franklin, Henry L. 9
Franklin's Corps Hospital, Hagerstown, Maryland 100
Fredericksburg, Marye's Heights, Virginia, Battle of 14
Fredericksburg, Virginia, Battle of 150, 151, 192, 194
Fremont, Illinois 85
Frese (Freese, Freeze), Frederick 147
Fritzmaurice, John 105
Fuld #138A 23
Fuld #142 23
Fuld #278 23
Fuld #279 23
Fuld #280 23
Fuld #280A 23

Fuld #339, 341, 342, 342A and 345 23
Fuld brothers 22

Gaines' Mill, Virginia, Battle of 82
General Order No. 33 6
General Order No. 204 197
George, Jacob P. 121
Georgia units: 2nd Infantry 194; 20th Infantry 194
German silver 25, 197
Germantown, Pennsylvania 115
Germany 126
Gettysburg, Pennsylvania, Battle of 25, 162
Giesbers, John 146
Gifford, Obed H. 18
Glasgow, Scotland 114
Gold gilding 25, 158
Gomersall, Wm. H. (sutler) 100
Graham, William 15, 101
Grand Rapids, Michigan 84
Grant, Lt. Gen. Ulysses S. 15
Gray, Stephen O. 89
Greek Cross (Sixth Corps Badge) 197
Green, Jacob 8
Green Castle, Pennsylvania 139
Greencastle, Scotland 106
Griffin, Brig. Gen. Simon 195
Grover, Stephen A. 143
Groveton, Virginia 192
Guigher, John I. 124

Hamilton, New York 111
Hammond U.S.A. General Hospital, Point Lookout Maryland 145
Hamson, Edward 142
Hancock, Maj. Gen. Winfield Scott 201
Hanson, C.H. 155
Harewood Hospital 110
Harper's Weekly magazine 16, 190
Harrisburg, Pennsylvania 116, 119, 120, 121, 134
Hartzog, Rich 201
Harvard Business School, Massachusetts 199
Hayward, W.A. 16, 198
Hennessey, John 192
Henry, Hall 124
Henry, Jesse 132

Henry Hill, Virginia, Battle at 5
Herrford, Maryland 81
Hilton, John J. 18
Hiserville, New Jersey 92
Holmes, Orrin 111
Hooker, Maj. Gen. Joseph 9, 16, 197
Hopkins, Ebenezer 144
Horter, Charles D. 23
Hotel tags 152
Huddersfield, England 137
Hull, Hiram 99, 151
Hurst, Robert H. 102

Identification badge 7, 8
Identification disc style 152
Identification disc sub-styles: All Seeing Eye (20A) 75; Banks (7A) 15, 59; Battles (4A) 15, 49; Battles (4B) 15, 50, 202; Battles (4C) 15, 21, 22, 51, 202, 203; Battles (4D) 15, 52; Double Blank (16A) 14, 70; Eagle (5A) 9, 10, 12, 21, 22, 23, 53, 155, 193, 195, 200, 203; Eagle (5B) 15, 22, 54; Eagle (5C) 15, 22, 55; Eagle (5D) 15, 22, 56, 199; Eagle (5E) 57; Engineer (13A) 67; Grant (9A) 15, 61; Grant (9B) 15, 62; Grant (9C) 15, 63; Hooker (11A) 15, 65; Lady Liberty (15A) 69, 202; Lincoln (6A) 15, 20, 22, 58; McClellan (1A) 13, 20, 23, 26, 155, 193, 200; McClellan (1B) 13, 20, 27, 194, 200; McClellan (1C) 15, 23, 28; McClellan (1D) 23, 29; McClellan (1E) 30, 200, 202; McClellan (1F) 15, 31, 200, 202; McClellan (1G) 15, 32, 200, 202; McClellan (1H) 33, 200, 202; McClellan (1I) 15, 34, 155, 202; McClellan (1J) 15, 35, 155, 200, 202; McClellan (1K) 15, 36, 155, 200, 202; McClellan (1L) 37, 200; McClellan (1Ma) 38, 200; McClellan (1Mb) 39; New York (14A) 68; Rhode Island (19A) 74; Scott (12A) 9, 66; Sherman (17A) 71; Sherman (17B) 72; Shield (2A) 12, 23, 40, 200, 202; Shield (2B) 23, 41, 199; Shield (2C) 15, 42, 200; Shield (2D) 43; Sigel (10A) 15, 64; Union (8A) 15, 60; Union League (18A) 73; Washington (3A) 10, 12, 13, 19, 21, 22, 44, 158, 159, 161, 197, 200, 201, 202, 203; Washington (3B) 13, 45; Washington (3C) 13, 46, 201, 202; Washington (3D) 47, 202; Washington (3E) 21, 48
Impact transfer method 158
Indiana units: 9th Infantry 15, 78; 20th Infantry 78, 79
Ingram, Pennsylvania 119
Iowa 198
Iowa units: 11th Infantry 14; 13th Infantry 13
Ireland 93, 114, 129

Jackson, Lt. Gen. Thomas J. 5
Janvrin, Joshus 91
Jenkins, John W. 13, 122, 192, 194, 204
Johnson, Col. Bradley T. (C.S.A.) 192
Johnson, John 98
Johnson, W.M. (sutler) 193
Johnston, Joseph E. 5
Johnston, New York 111
Jones, Yeaman 7
Jones' House, Virginia 195
Juniata County, Pennsylvania 123

Kearny, Phillip 9
Keene, New Hampshire 86, 189
Kennedy, John 14
Kentucky units: 12th Infantry(?) 202
Kerr (Carr), Christian 126
Kirby, James 87, 192, 193, 194, 195
Knoxville, Tennessee 192
Kunkle, Reuben S. 127

Lafayette, Indiana 78, 79
Lancaster, Pennsylvania 12, 104, 116
Lead 25, 197, 203
Letter punch defect, McClellan (1D) 160
Letter punch, gang 201
Letter punch, gang of battles 151
Letter punch, individual of battles 151
Letter punch kit 159, 198
Letter styles: Eagle (5A) 160; McClellan (1A) 160; Shield (2A) 160; Washington (3A) 160
Lettering 159
Lewis, Isaac E. 116
Lewis, Joseph 110
Lewis, Leigh, Jr. 198
Lincoln, Abraham 12, 195
Livingston, William 94
Longford, Ireland 108
Lord, Francis 19, 20, 139
Louisiana units: 20th Infantry C.S.A. 147
Lovett, George H. 21, 47, 48
Lovett, John D. 74, 201
Lovett, Robert, Jr. 21, 23, 202
Lown, Henry 113, 152
Lucas, Andrew 124
Luggage tag 149, 152, 153
Lynnfield, Massachusetts 82

Maine units: 17th Infantry 80
Manassas, First Battle of 7, 9, 12, 14
Manassas, Second Battle of 14, 25, 189, 192, 194, 195
Mansion House Alexandria Hospital, Virginia 80
Maryland units: 2nd Infantry 81, 189, 192, 194, 195
Marston, Archibald 146
Massachusetts units: 9th Infantry 82; 10th Infantry 19; 19th Infantry 82, 150
Matthews, George A. 108
Mayo, Col. Joseph 202
Mays, Samuel 77
McCabe, Walter 138
McCleary (McClary), Samuel E. 118
McClellan, Maj. Gen. George B. 6, 9, 10, 13, 23
McCreery, William H. 134
McDevett, Neal 15, 93
McDowell, Irvin 5
McInnes, James 13, 113
McIntyre, James 125
Medal 9, 18, 24
Medal of Honor 21
Memorandum from Prisoner of War Records 195
Memorial Brass I.D. Manufacturing Co. 153
Merchant tokens 201
Merriam, Joseph H. 20, 21
Metal detector 9
Mexican War 6
Michigan units: 2nd Infantry 83; 3rd Infantry 84; 17th Infantry 83

Milesburg, Pennsylvania 124
Millar, Percy 18
Minersville, Pennsylvania 122
Minted counterfeit coins 158
Mississippi units: 2nd Infantry 202
Mitchell, John 13, 109
Monroe County, Pennsylvania 127
Montpelier, Vermont 144
Montrose, Pennsylvania 126, 137
Moore, George, Jr. 106
Morgan, George W. 115
Morgan, Paul 86
Morrison, James A. 116
Myers, Henry 8
Myers, Norris 7

Nagle, Col. James 189, 190, 192, 195
New Hampshire 196
New Hampshire units: 2nd Infantry 85, 86; 5th Infantry 86; 6th Infantry 87, 189, 192, 193, 194, 195; 9th Infantry 88, 189, 190, 192, 195; 11th Infantry 195; 12th Infantry 89, 90; 14th Infantry 20, 85, 91; 15th Infantry 91
New Jersey 102
New Jersey units: 5th Infantry 92; 10th Infantry 92; 11th Infantry 93
New York 77
New York City 14, 20, 21, 23, 96, 101, 102, 103, 105, 109, 113
New York units: 2nd Heavy Artillery 112; 2nd Veteran Cavalry 94; 4th Independent Artillery 13, 105; 4th Heavy Artillery 113, 152; 9th Infantry 10; 16th Heavy Artillery 16th 113, 152; 22nd Infantry 94; 25th Infantry 95, 151; 30th Infantry 96, 152; 31st Infantry 96, 150; 33rd Infantry 97, 98; 34th Infantry 98; 37th Infantry 99, 151; 40th Infantry 106, 108; 43rd Infantry 100; 56th Infantry 15, 101; 57th Infantry 101; 61st Infantry 102, 103; 70th Infantry 12, 104; 71st Militia Infantry 13; 72nd Infantry 13, 104; 74th Infantry 105; 76th Infantry 106; 80th Infantry 15; 87th Infantry 107, 108; 101st Infantry 109; 103rd Infantry 13; 145th Infantry 109; 147th Infantry 110; 153rd Infantry 111; 157th Infantry 111; 179th Infantry 99
Newport, Pennsylvania 122
Newville, Pennsylvania 141
Nickel 25, 197
99th Company, 2nd Battalion Invalid Corps 86
Ninth Corps 189, 192, 195
North Anna River, Virginia, Battle of 192
North Carolina units: 1st Infantry 202
Northrop, Edward 96, 150
Numismatic Guaranty Corporation of America 149, 201

O'Blenis, Henry 78
Occupations: ambrotypist 139; ambulance driver 95; blacksmith 102; boatman 93; brakeman 124; carpenter 98, 130; clerk 80, 111; coach trimmer 116; cook 100, 102; engineer 132; farmer 77, 78, 79, 84, 86, 89, 90, 92, 96, 99, 100, 110, 113, 119, 121, 125, 126, 129, 145, 146; fuller 137; glove maker 111; guard train 141; harness maker 118; hostler 84; iron molder 114; laborer 81, 86, 104, 105, 114, 120, 131, 141, 192; machinist 116, 132; mariner 126; mechanic 76; miner 122, 194; musician 133; nurse 145; nurse/cook 100; orderly 114, 141; painter 101, 134; paper hanger 133; paper maker 146; plasterer 139; principal musician 89; roller 124; sailor 148; salesman 94; shoemaker 82; stage driver 77; tailor 143; teacher 134; teamster 85, 110, 121, 125; wagoner 85; wardmaster 145; wheelwright 140
Ohio units: 5th Infantry 13, 113; 61st Infantry 114
119th Regiment Colored Troops 115
O'Neal, Patrick 112
Online auctions 163
Online sales 162
Osborne, Henry 16
Overland Campaign, Virginia, Battle of 195

Paine, A. 12
Pantucket, Massachusetts 131
Paper identification tag 7
Paris, Kentucky 195
Patriotic token 22, 199, 200
Peck & Snyder 153
Pender, Dorsey 153, 202
Pender disc 153
Pennsylvania 25
Pennsylvania units: 2nd Cavalry 140; 2nd Heavy Artillery 39, 140; 3rd Cavalry 141; 3rd Reserve Volunteer Corps Infantry 115, 116; 5th Cavalry 141; 5th Reserve Volunteer Corps Infantry 116, 117; 11th Reserve Volunteer Corps Infantry 118; 12th Cavalry 201; 12th Reserve Volunteer Corps Infantry 119; 13th Infantry 130; 14th Infantry 138; 19th Cavalry 142; 20th Emergency Volunteer Infantry 142; 26th Infantry 119; 31st Infantry 128; 45th Infantry 120, 121; 46th Infantry 136; 47th Infantry 121, 122; 48th Infantry 13, 122, 189, 190, 194; 49th Infantry 123, 124; 53rd Infantry 124; 56th Infantry 124, 125, 126; 67th Infantry 127; 69th Infantry 128; 71st Infantry 131; 82nd Infantry 128; 83rd Infantry 129; 100th Infantry 129; 102nd Infantry 130; 104th Infantry 138; 110th Infantry 131; 115th Infantry 131; 119th Infantry 132, 133; 121st Infantry 133, 134; 128th Infantry 121; 135th Infantry 134; 137th Infantry 134; 143rd Infantry 136; 148th Infantry 135; 149th Infantry 136; 150th Infantry 136; 151st Infantry 137; 155th Infantry 138; 184th Infantry 124; 191st Infantry 117; Battery D Independent Artillery 138
Percy, E.L. 22
Perry, Charles L. 88, 192, 195, 204
Perry, D. 112
Petersburg, Virginia, Battle of 192, 194
Pewter 25, 197
Philadelphia, Pennsylvania 21, 23, 115, 119, 128, 132, 133, 134, 136, 142

Index

Phillips, H. 13
Phillips, James H. 77
Phillips, Stanley S. 15, 16, 197
Pierson, James H. 111
Pine Creek, Pennsylvania 140
Pittsburgh, Pennsylvania 130
Pittsfield, New Hampshire 90
Pittston, Pennsylvania 140
Plastic molds 158
Point Lookout, Maryland Hospital 89
Political medals 152
Poplar Spring Church, Virginia 195
Portland, Maine 80
Post, Reuben M. 136
Potomac River 12
Preble County, Ohio 78
Preprinted identification message 7
Price 162
Price, Isaac A. 140
Private auctions 163
Professional Coin Grading Service 202
Provenance 150
Providence, Rhode Island 103, 143, 198

Railroad Cut, Virginia, Battle at 189
Railroad iron 194
Raritan, New Jersey 107
Rarity: common 25, 53, 200; extremely rare 25, 32, 33, 34, 35, 36, 37, 38, 39, 43, 48, 49, 52, 56, 57, 59, 60, 61, 62, 63, 65, 66, 67, 68, 69, 71, 72, 73, 74, 75; rare 25, 27, 29, 41, 42, 45, 50, 51, 55, 58; scarce 25, 26, 28, 40, 44, 46; very rare 25, 30, 31, 47, 54, 64, 70
Reading, Pennsylvania 138
Reed, Charles Wellington 198
Reed, Francis 126
Re-enactors 150, 155
Relic hunters 9
Reproduction identification discs: Eagle (5A) 153, 155, 157; Identification Discs 155; McClellan (1A) 156; McClellan (1K) 156; Shield (2A) 156; Shield (2B) 155
Reproductions and fakes 152
Revolutionary War 6
Rheumatism 14
Rhode Island units: 1st Light Artillery 15, 143; 3rd Cavalry 143; 7th Infantry 192; 11th Infantry 143; 12th Infantry 192
Rice, Henry T. 121
Richmond, Kentucky 195
Ridley, George 109
Rim 157
Rockhill, Pennsylvania 126
Rohrback Bridge, Maryland 189
Rollins, David 144
Rulau, Russell 202

St. Albans, Vermont 145
St. Louis, Missouri 78
Salisbury, North Carolina 196
Sanborn, Theodore 90
Sanbornton, New Hampshire 90
Sand molds 158
Sanitary Commission 9
Savage Station, Virginia, Battle of 14
Schenkman, David E. 199
Schuylkill County, Pennsylvania 189
Scotland 79
Scott, John 76
Scott, Maj. Gen. Winfield 9
Scovill Manufacturing Company 22, 23, 199
Scovill's daguerreotype materials 22
Seabrook, New Hampshire 91
Senior, Thomas R. 119
Serif 160, 161, 164
"7 Days" 151
Shaffer, John 120
Sharp Shooter Division Headquarters 146
Sharpsburg, Maryland 190
Sheehan, William J. 134
Shenandoah Valley, Virginia 20
Sherman, Maj. Gen. William T. 201
Shield 9
Shoop, James M. 141
Sigfried, Lt. Col. 190
Silver 25, 197
Silver wash 25, 158
Sixth Corps 13
Slabs 149
Slough General Hospital, Alexandria, Virginia 85
Smith, Frederick B. 20, 23, 26, 58
Soil: acidity 159; composition 159; moisture 159
Soldier records 150
Soldier's identification medal 155
South Mountain, Maryland, Battle of 189, 194, 195
Spanish-American War 202
Spark erosion method 158
Spotsylvania, Virginia, Battle of 15, 192, 198
Spotsylvania Court House, Virginia 194
Springfield, Illinois 77
Stanton, Edwin M. 14
Stanton, Robert 100
Staten Island, New York 104
Steele, George 96, 152
Stewart, William 129
Stone Bridge, 1st Manassas, Virginia, Battle at 6
Stone U.S.A. Hospital 110
Store cards 22, 199, 200
Stratton, Theodore F. 128
Strike 157
Sullivan, Owen 114
Sunbury, Pennsylvania 118
Surface appearance 156
Sutler 9, 18, 19, 20, 24, 151, 155, 160, 163, 193, 194, 198, 201, 204
Sweatt, Charles L. 90
Swiler, John F. 135

Tarnish remover 203
Taylor, Thomas 202
Taylor, William C. 102
10th Regiment Veteran Reserve Corps 91
Texas units: 8th Cavalry 202
39th Company, 1st Battalion Invalid Corps 79
Thomas, Reese J. 130
Thompsonville, Connecticut 76
Thurstop, James F. 18
Tin 25, 197
Titus, Lt. H.B. 192
Toe tags 7
Toms, Joseph H. 101
Trac(e)y, Thomas 81, 192, 193, 204
Troy, New York 94, 96
Turnbridge, Vermont 144

Union shield 10
United States Military History Institute, Carlisle, Pennsylvania 198
United States Mint 163
United States National Archives 150, 201

United States National Park Service 201
United States Signal Corps 134
United States units: 1st Volunteer Infantry 147
United States War Department 14
University of Connecticut 199
U.S.A General Hospital, Portsmouth Grove, Rhode Island 79

Vairin, A.L.P. 202
Vermont units: 2nd Infantry 9, 144; 3rd Infantry 144; 5th Infantry 145; 6th Infantry 146; 7th Infantry 145
Vicksburg, Mississippi 192
Virgil, New York 106
Virginia Units: J.S. Brown's Artillery Battery 194; 3rd Infantry 202

Wakarusa, Indiana 78
Wakefield, Pennsylvania 129
War medal 201
"War of 1861" 161, 164, 200
Ward, Harrison B. 140
Warrenton Turnpike, Virginia 5
Washington, George 10
Washington, James B. 153, 202
Washington, D.C. 5, 118, 131, 189, 195
Waterbury, Connecticut 21
Wayne County, Pennsylvania 125
Welsh, Isacc 95, 151
West Lebanon, New Hampshire 86
West Point, Virginia, May 7, 1862 150
Westerly, Rhode Island 112
White metal 10, 25, 197

Whitehill, Andrew L. 135
Wilderness, Virginia, Battle of 15, 198
Williams, Robert 98
Wilmington, Delaware 77
Wilmont, Albert E. 103
Wilson, Joseph S. 142
Winchester, Second Battle of 201
Wisconsin units: 6th Infantry 146; 11th Infantry 23
Wood, Jonathon 136
Word configuration 159, 160; Eagle (5A) 160; McClellan (1A) 160; Shield (2A) 160
Wright, John 114

Yeomans, R.S. 20

Zeisert, S. 12
Zinc 25

www.ingramcontent.com/pod-product-compliance
Ingram Content Group UK Ltd.
Pitfield, Milton Keynes, MK11 3LW, UK
UKHW011331020225
454438UK00011B/125